I0011118

PLC Hardware and Programming

Multi-Platform

February 2022

Revision 1.2

Course Manual

By

Frank Lamb

www.automationllc.com

Overview

This course approaches PLC training from a generic viewpoint. Most PLC platforms have many things in common; before beginning the study of a particular brand of PLC, it is important to learn the things that are common to *all* platforms. This class does this, pointing out some of the exceptions and different ways of doing things along the way. Resources used in the preparation of this course include information from many of the major PLC manufacturers; software examples are primarily drawn from Allen-Bradley RSLogix5000 and Siemens Step 7.

Generic Modules:

Introduction

What is a PLC, or Programmable Logic Controller?

A **PLC**, Programmable Logic Controller or Programmable Controller, is a digital computer used to control electromechanical processes, usually in an industrial environment. It performs both discrete and continuous control functions and differs from a typical computer in several important ways:

1) It has **Physical I/O**; electrical inputs and outputs bring real world information into the system and control real world devices based on that information.
2) It is **Deterministic**; it processes information and reacts to it within defined time limits.
3) It is often **Modular**; it can have I/O modules, communication modules or other special purpose modules added to it for expansion.
4) It is programmed using several defined **Languages**. Some languages allow the program to be changed while the machine or system being controlled is still running.
5) Software and Hardware are **Platform Specific**; components and programming software usually can't be used between different manufacturers.
6) It is **Rugged** and designed for use in industrial environments.

Unlike computers, PLCs are made to run 24 hours a day, 7 days a week and are able to resist harsh physical and electrical environments.

According to a *Control Engineering* magazine poll in 2012, major applications for PLCs include machine control (87%), process control (58%), motion control (40%), batch control (26%), diagnostic (18%), and other (3%)." (Results don't add up to 100% because a single control system generally has multiple applications.)

Allen Bradley ControlLogix

Siemens S7-1500

Mitsubishi Q Series

Koyo DL405

Omron CJ1

GE 9030

Modicon Premium

History and Evolution

While PLCs have only been in existence for fewer than 50 years (as of 2016), the evolution from computing machines into their current form is interesting and illuminating. This section examines their history and parallel development with computers. To understand the history of programmable controllers, it is useful to examine where many of its elements came from.

The Babbage Analytical Engine

Even before the use of electrical devices to solve mathematical and logical problems, a mathematician and inventor from England, Charles Babbage, had an idea for a mechanical device to calculate astronomical and mathematical tables. In 1823, he was awarded £1700 by the British government to begin work on his "Difference Engine", but after spending more than £17,000 over 15 years or so, no working device had been built. The government viewed this as a total failure on the mechanical side, though they did consider the economically produced tables themselves worthwhile.

In 1837, he proposed a mechanical general purpose computer to solve arithmetic problems. The "Analytical Engine" was designed to solve general mathematical polynomial equations using a system of gears. The work Babbage did on the Analytical Engine made the original Difference Engine idea obsolete, at least in his mind. Because of conflicts with his chief engineer and what he considered inadequate funding, he was never able to build a working device.

Many of the concepts that came out of Babbage's designs laid the groundwork for future computers and processors. For instance, his designs incorporated an Arithmetic Logic Unit, Control Flow (conditional If-Then branching and Loops), and integrated memory in the form of gear positions. This structure is similar to the electronic version of the computer.

If the Analytical Engine had been built, it would have been digital and programmable. It also would have been "**Turing-complete**", a term that when applied to programming languages means that it has conditional branching (*e.g.*, "if" and "goto" statements, or a "branch if zero" instruction) and the ability to change an arbitrary amount of memory.

The original engine would, however, have been very slow. In *Sketch of the Analytical Engine*, Luigi Federico Menabrea reported "Mr. Babbage believes he can, by his engine, form the product of two numbers, each containing twenty figures, in three minutes".

Relay Logic

Before the development of computers, automated control of machinery was accomplished by wiring relays and other devices together. Electromechanical relays and switches could be used to control pumps, heaters and motors when connected in a "relay rack". This method was expensive, took up a lot of space and was difficult and time consuming when it came to making changes.

Methods of designing control logic using hand-drafted "ladder diagrams" had been used since the advent of electrical systems in the 1800s. Ladder diagrams were named after their resemblance to the rungs of a ladder, with the left side being the line, or energizing side (L1) and the right side being the neutral, or L2 side.

The electromechanical relay itself, invented in the early 1800s, was not used widely until telegraphs and telephone switching circuits required them in the late 19th century. In the early 1900s, inventors realized that electrical relay circuits could be used to direct a series of mathematical calculations automatically. Simple relays were fairly inexpensive by the 1930s, but it was still less costly to build "adding machines" using mechanical cams and gears.

For more complex math functions, relay circuits were more flexible than mechanical systems. They could be arranged in rows and columns on a rack and connected together with wires, according to what one wanted the circuit to do.

Electromechanical Computers

In 1937, the concept of a general purpose computing machine was presented to IBM by Howard Aiken. The project was approved in 1939 after a feasibility study and completed in 1944. The first version, the Mark I, was shipped to Harvard and used to determine whether implosion was a viable choice to detonate the atomic bomb that was used a year later. Interestingly, Aiken's ideas were clarified after studying Babbage's work one hundred years earlier. The Mark I was also used to compute and print the same kinds of mathematical tables that were Charles Babbage's original goal.

The Mark I was built using switches, relays, rotating shafts and clutches. It weighed about 10,000 lbs. (4500kg) and was 51 feet long. It used 765,000 components and 500 miles of wire. There were 3500 multipole relays with 35,000 sets of contacts, 2225 counters and tiers of 72 adding machines, each of which could do math with 23 significant digits. The basic calculating units were synchronized mechanically by a 5 horsepower (4kW) electric motor.

Data could be entered by using 60 sets of 24 switches, but could also read instructions from a 24 channel punched paper tape.

The Harvard Mark I was followed by the Harvard Mark II in 1947, the Harvard Mark III/ADEC in 1949, and the Harvard Mark IV in 1952. While the Mark II was an improvement over the Mark I, it was still based on electromechanical relays. The Mark III used vacuum tubes and crystal diodes, but still included rotating mechanical drums for storage and relays for transferring data between drums.

The Mark IV was all electronic, replacing the drums with magnetic core memory. The Mark II, III and IV were all sold to the military (U.S. Navy and Air Force). The Mark I remained at Harvard and was retired in 1959. An early picture of the computer displays the name "Aiken-IBM Automatic Sequence Controlled Calculator Mark I".

The First Electronic Computers

The first electronic general purpose computer was designed to calculate artillery firing tables for the U.S. Army but was also used to study the feasibility of a thermonuclear weapon. It was originally announced in 1946 and called a "Giant Brain" by the press. It was designed to be about 1000 times faster than an electromechanical computer.

This computer was called ENIAC, an acronym for Electronic Numerical Integrator and Computer. It was operational from 1946-1955, but improvements were made in 1948 (a read-only stored programming mechanism was added), 1952 (High Speed Shifter added) and 1953 (100-word BCD Core Memory added). These improvements increased the speed and capabilities of the computer quite a bit.

ENIAC contained 17,468 vacuum tubes, 7200 crystal diodes, 1500 relays, 70,000 resistors, 10,000 capacitors and about 5,000,000 hand-soldered connections. It was 100 feet long and weighed about 27 tons (54,000 lbs). It required 150kW of electricity; the rumor was that when it was switched on, the lights in Philadelphia dimmed.

Input was possible using an IBM card reader and output was via a card punch or by using an IBM accounting machine. It was modular, consisting of individual panels to perform different functions. Some panels were accumulators that performed math functions, other units included the Initiating Unit which started and stopped the machine, the Cycling Unit which synchronized the other units, the Master Programmer which controlled loop sequencing, the punch card reader, the printer, the Constant Transmitter and 3 Function Tables, which were programmed using switches.

ENIAC used octal-base radio tubes, which were common. Several tubes burned out daily, making it non-functional about half the time. Higher reliability tubes appeared in 1948, reducing downtime.

Mapping a mathematical problem into ENIAC could take weeks; it was quite complex. Manipulations of switches and cables followed by verification and debugging was required to execute the program step by step. There were six female programmers who not only input the data but also debugged problems by crawling inside the machine to find bad solder joints and tubes.

The picture to the left shows the back of one section of ENIAC, full of vacuum tubes.

Near the end of World War II, the U.S. Navy had approached MIT about creating a flight simulator for bomber pilots. The project was funded under the name "Project Whirlwind".

After seeing a demonstration of ENIAC, an engineer at MIT suggested that a digital computer was the solution. By 1947, a high-speed stored-program had been built. Most of the previous computers at this time operated in bit-serial mode, using single bit arithmetic and feeding in large words of 48 or 60 bits in length. This method was not fast enough for the simulation task, so the Whirlwind computer included 16 math units that operated on 16-bit words in bit-parallel mode. This made Whirlwind sixteen times faster than other machines of its era. This legacy has been handed down to the CPUs of today which almost all use this bit-parallel system to do arithmetic.

Official Whirlwind construction began in 1948 and took three years to build. It began operation on April 20, 1951. The design used approximately 5,000 vacuum tubes.

Early Computer Memory

In the first electromechanical and electronic computers, memory took the form of mechanisms, latched relays or vacuum tubes held in a fixed state. This method was slow and not very flexible.

The original Whirlwind computer design called for 2048 words (2K) of 16 bits each for random-access storage. In 1949, when the design was being done there were only two memory technologies that could hold this much data; **mercury delay lines** and **electrostatic storage**.

A mercury delay line was a complex system that consisted of a long tube filled with mercury, a mechanical transducer on one end and a microphone on the other. Pulses were sent in at the transducer end, detected by the microphone, reshaped and sent back through the delay line, much like a spring reverb unit used in audio processing. The delay line operated at the speed of sound, which was very slow even by the standards of computers in that era. Whirlwind designers discarded the delay line as a possible memory resource due to its speed and complexity.

The alternative form of memory, electrostatic memory tubes, used a cathode ray tube similar to an early TV picture tube or oscilloscope tube. The most popular type then was known as the Williams tube, developed in England, but this design was incompatible with the Whirlwind specifications, so the designers chose a different model.

After spending many months testing the system, it was determined that the electrostatic tubes were too slow for the project requirements, and a suitable replacement was sought. The leader of the project was an engineer named Jay Forrester. He had come across an advertisement for a new magnetic material that he recognized as a possible data storage medium and set up a workbench in the corner of the lab to evaluate the material. During several months at the end of 1949, he had invented the basics of magnetic core memory and demonstrated its feasibility.

After two more years of work, the design team had built a core plane that was made up of 32 x 32, or 1024 individual cores, holding 1024 bits of data. Two additional core planes were built later, increasing the total memory of the system to 3072 bits.

Core memory used tiny magnetic rings with wires threaded through them to read and write information. The cores can be magnetized in two different ways, clockwise or counterclockwise, the bit stored is a zero or a one based on the magnetization direction.

Wires are arranged so that individual cores could be set to a one or zero by changing its magnetization, but reading the core caused it to be reset to zero, erasing it. This is known as destructive readout. This problem was solved in 1951 by An Wang, who invented a write-after-read cycle that used a one dimensional shift register of cores, essentially using two cores to store each bit.

When not being read or written to, the core retains its value, even when power is removed. This makes them **nonvolatile**.

By using smaller rings and wires, the memory density slowly increased; by the late 1960s a density of 32 kilobits per cubic meter was typical. Costs for manufacturing these core units declined from $1 per bit to 1 cent per bit by 1970.

The first semiconductor-based memory (Semiconductor Random Access Memory or SRAM) was introduced in the late 1960s and began to erode the core market. In 1972, the first successful DRAM (Dynamic Random Access Memory) Intel 1103 was brought to market at 1 cent per bit. Improvements in semiconductor manufacturing led to rapid increases in capacity and further decreases in price, driving core from the market by 1974.

The Evolution of Personal Computers

Computer technology improved, making computers more powerful and smaller through the 1960s and 1970s, however they were still too large and expensive for individuals or even small companies to own one. Requests for computer use at universities and larger businesses had to be filtered through an operating staff or time-shared.

By 1972, electronic calculating devices began to be available to individuals and small businesses. After development of the microprocessor, individual personal computers were low enough in cost that they could be purchased, though they were often sold as a kit and of interest mostly to hobbyists and technicians.

By the 1980s, several commercially available computers were being sold, including the TRS-80 (released 1977 by Radio Shack), Atari's 400 and 800 (late 1970s), and the Commodore 64 (1982).

IBM introduced its 5150 model in 1981, and through the 1980s the term "PC" began to refer to desktop computers compatible with IBM's PC products. In 1970, IBM was estimated to have had 60% of the computer market; by 1980 this had declined to 32%. By 1984, it was estimated in a *Fortune* magazine survey that 56% of American companies with personal computers used IBM PCs, compared to 16% for Apple. A 1983 study of corporate customers found that two thirds of large customers standardizing on one computer chose the IBM PC, compared to 9% for Apple. Through the 1980s, the only product that kept a significant market share without being compatible with IBM personal computers was the Apple Macintosh family.

The 5150 was based on the Intel 8088 and was their first open architecture product, allowing competitive companies to create peripheral products for their computer, including software. IBM didn't market its own software until 1984, instead relying on licensing programs like

Microsoft's BASIC. The PC/AT was introduced in 1984 with an Intel 80286 CPU and originally running at 6MHz clock speed. This was followed by the XT 286 in 1986, which was the first personal computer designed to support multitasking.

Almost immediately after the IBM PC was released, rumors began circulating of compatible computers created without IBM's approval. In June 1982, Columbia Data Products introduced the first IBM-PC compatible computer, followed by Compaq in November of 1982. Other manufacturers reverse engineered the BIOS to make non-patent-infringing copies of the operating system.

Early IBM PC compatibles used the same computer bus as the original PC and AT. This was later named the Industry Standard Architecture by manufacturers of compatible computers. In June 1983, PC Magazine defined a "PC Clone" as "a computer that can accommodate the user who takes a disk home from an IBM PC, walks across the room and plugs it into the 'foreign' machine".

Other manufacturers such as Tandy (Radio Shack), Hewlett-Packard, Digital Equipment Corporation and Texas Instruments introduced computers that were compatible with Microsoft's "MS-DOS", though these were not always completely hardware and software compatible with the IBM PC itself. As more companies began manufacturing these "clone" computers, IBM began to lose market share again. Into the 1990s, compatibility became more of a matter of software than hardware as Microsoft created the Windows series of operating systems and most software companies concentrated on Windows compatible products. IBM finally exited the personal computing market with the sale of its consumer PC division to Lenovo in 2005.

In the 2000s, Hewlett Packard and Dell have become the largest American PC manufacturers. Processors have become more powerful every year and the footprint has shrunk, with laptops and tablets taking up much of the market. Major foreign manufacturers including Acer, Lenovo, Sony and Toshiba are also major competitors in the market, and prices have been lowered extensively to where often software costs more than the hardware it runs on.

Tablets with detachable keyboards and laptops with touchscreens have further blurred the lines between handheld devices and full computer platforms. In addition, connectivity and remote servers (the "Cloud") have distributed software services between in-house computers and service-based software.

Birth of the Programmable Controller

In 1968, a group of engineers at General Motors presented a paper at the Westinghouse conference detailing the problems they were having with reliability and documentation of the machines at their plant. One of the engineers, Bill Stone, also presented design criteria for a "standard machine controller".

The criteria stated that the design would need to eliminate costly scrapping of assembly-line relays during model changeovers and replace unreliable electromechanical relays. It also needed to:

- Extend the advantages of static circuits to 90% of the machines in the plant.
- Reduce machine downtime related to controls problems, be easily maintained and programmed in line with already accepted relay ladder logic.
- Provide for future expansion. It had to be modular to allow for easy exchange of components and expandability.
- It had to work in an industrial environment with dirt, moisture, electromagnetism and vibration.
- Include full logic capabilities, except for data reduction functions.

These specifications, along with a proposal request to build a prototype, were given to four controls builders:

- Allen-Bradley, by way of Michigan-based Information Instruments, Inc.
- Digital Equipment Corporation (DEC)
- Century Detroit
- Bedford Associates

The DEC team brought a "mini-computer" to GM, which was rejected. A lack of static memory was one of the major reasons.

Allen-Bradley was a major manufacturer of relays, rheostats and motor controls. Even though this new idea would compete with one of its core businesses, electromechanical relays, they went from prototype to a production unit in 5 months. The first attempt was the Program Data Quantizer, or PDQ-II. This was judged to be too complex and difficult to program and was quite large. The next attempt was the Programmable Matrix Controller, or PMC. Though smaller and easier to program, this was still not sufficient for GM.

At the time of GM's design criteria, Bedford Associates was already working on a design. Its system was modular and rugged, it used no interrupts for processing and mapped directly into memory. Since this was the 84th project for the company, they named this unit the 084. The project team included Richard Morley, Mike Greenberg, Jonas Landau, George Schwenk and Tom Boissevain. After obtaining funding, the team formed a new company called Modicon, an acronym for **MO**dular **DI**gital **CON**troller.

From left to right: Dick Morley, Tom Boissevain, Modicon 084, George Schwenk, and Jonas Landau

The Modicon 084 was built ruggedly, with no on-off switch, no fans, and totally enclosed. Richard Morley explained, *"No fans were used, and outside air was not allowed to enter the system for fear of contamination and corrosion. Mentally, we had imagined the programmable controller being underneath a truck, in the open, and being driven around in Texas, in Alaska. Under those circumstances, we wanted it to survive. The other requirement was that it stood on a pole, helping run a utility or a microwave station which was not climate controlled, and not serviced at all"*.

In 1969, Bedford and Modicon demonstrated their 084 Programmable Controller to GM and won the contract. The controller consisted of three components: the processor board, the memory, and the logic solver board, which used a form of ladder logic to solve the algorithms.

According to Morley, the original machine only had 125 words of memory and did not need to run fast. In his interview with Howard Hendricks, he said *"You can imagine what happened! First, we immediately ran out of memory, and second, the machine was much too slow to perform any function anywhere near the relay response time. Relay response times exist on the order of 1/60th of a second, and the topology formed by many cabinets full of relays transformed to code is significantly more than 125 words. We expanded the memory to 1K and thence to 4K. At 4K, it stood the test of time for quite a while. Initially, marketing and memory sizes were sold in 1K, 2K, 3K, (?) and 4K. The 3K was obviously the 4K version with constrained address so that field expansion to 4K could easily be done."*

Allen-Bradley Bulletin 1774 PLC

Meanwhile, Allen-Bradley had gone back to the drawing board. By 1971, engineers Odo Struger and Ernst Dummermuth had begun to develop a new concept that improved on their PMC, Programmable Matrix Controller. This concept became the Bulletin 1774 PLC. Allen-Bradley named this the "Programmable Logic Controller"; the term later became the industrial standard when the acronym PC became associated with personal computers.

In 1972, Allen-Bradley also offered the first computer for use as a programming terminal. Other manufacturers in the 1970s and 1980s typically used dedicated programming terminals with (or without) a small screen. Instructions were entered as three or four letter mnemonics. As technology improved, these terminals were reduced in size to a hand-held device.

Mitsubishi F1-20P Programmer, 1980s

By the later years of the 1970s, several other companies had entered the PLC market, including General Electric, Square D, Omron and Siemens.

Modicon improved on the 084 in 1973 with the 184, which made them the early leader in the market. This was followed in 1975 with the 284 and 384 models. The 984 was produced in 1986 and remained a Modicon standard for many years. In a joint venture with AEG Schneider Automation, the Quantum series of controllers was released in 1994. In 1977, Modicon was bought by Gould Electronics, and later in 1997 by Schneider Electric, who still owns them today (2016).

PLC Improvements

The 1980s saw many new companies entering the PLC market. Japanese companies such as Mitsubishi and Omron entered the U.S. market as automotive manufacturing began using PLCs extensively in their manufacturing processes. Giants like Westinghouse, Cutler Hammer and Eaton created products, as well as machine tool manufacturers such as Giddings & Lewis. In 1980, the market was estimated at $80 million, and it had grown to a billion dollars worldwide by 1988.

As IBM-compatible personal computers became smaller and less expensive, companies began developing DOS-based software for use in programming. This allowed users to enter the program graphically. Rather than only seeing the alphanumeric characters of text commands, ladder logic could be visualized on a CRT monitor.

With the release of the Windows 3.0 operating system in the early 1990s, software had improved with colored graphics and multitasking. PC clones made computers less expensive, and the laptop computer all but replaced handheld programmers. Many companies began producing smaller, cheaper "brick" PLCs for simple applications. Allen-Bradley's Micrologix 1000 in 1995 competed with relatively unknown names, such as Eagle Signal's "Micro 190" and PLC Direct, which brand labeled Koyo PLCs from Japan.

Koyo had produced PLCs for Texas Instruments, Siemens, and GE since the 1980s. In 1994, they established a company in the U.S. that began marketing PLCs by mail order. Tim Hohmann founded PLC Direct in Atlanta as a joint venture, which was renamed Automation Direct in 1999.

Allen-Bradley remained the dominant brand in the U.S. during the 1990s. They had been bought by Rockwell in 1984 and spun off Rockwell Software in 1994 after purchasing ICOM, which had made a competing programming software product for the Allen-Bradley PLC. The SLC500 line, a smaller modular controller, was released in 1991, followed by the first MicroLogix product in 1995.

Siemens became the dominant player outside of the U.S. and Japan. The S5 controller, developed in 1979, had a large installed base in Europe through the 1980s. When the S7-200, S7-300 and S7-400 series were released in 1994, many companies began upgrading their existing platforms. Siemens was an early innovator in the use of User Defined Data Types (UDTs) and advanced programming using their version of Instruction List, known as STL (Statement List). They also allowed for use of re-useable code by defining local variables within subroutines, or functions.

In 1994, the International Electrotechnical Commission (IEC) began to define the languages that PLCs would be programmed in, data types, and other details pertinent to Programmable Controllers. IEC 61131-3 defined the rules manufacturers followed in order to standardize their products. Five languages were defined: Ladder (LD), Instruction List (IL), Function Block Diagram (FBD), Structured Text (ST), and Sequential Function Charts (SFC).

Mitsubishi gained the largest market share in Japan and much of Asia, while Omron saw gains worldwide.

By the 2000s, PLCs had become much more powerful and began gaining traction in process control, which had long been the domain of DCS (Distributed Control Systems). With the ability to use I/O networks such as DeviceNet, Profibus and Ethernet, these more powerful platforms became known as "Programmable Automation Controllers", or PACs. With improved memory, higher speed processors and the ability to control thousands of analog and digital points at once, PACs could control large chemical processing plants, wastewater treatment and pipelines.

Multi-axis motion control also began to be integrated into PACs in the early 2000s. Allen-Bradley, Siemens, Modicon and Mitsubishi all have integrated controllers that can operate independently of the central CPU. Multiple or redundant CPUs can also be used within the same rack. Variable Frequency Drives (VFDs) and robots often now contain microprocessors that can be programmed in ladder logic. Hybrid HMI touchscreen controllers have also become common.

Today's landscape includes more than 20 PLC manufacturers with international markets, with about 15 that have a 1% or more market share each. Open platforms have appeared allowing smaller manufacturers to offer their own PC-based or board-level controllers that program in ladder or other IEC 61131-3 languages. Companies such as Codesys now provide a platform for some of the major PLC manufacturers such as Modicon, ABB and Bosch.

With higher speed Ethernet-based communication and control networks, systems have become more distributed, with microprocessors in "smart nodes" of I/O to detect errors and perform autonomous logic and monitoring tasks. As of 2016, the PLC market seems to be converging on Ethernet/IP based control networks.

PLC Timeline

PLC Timeline 1968 - 2016							
Event	Event	Network	Product	Company	Acquisition	PLC	Platform
Year	Events	Modicon	AB	GE	Omron	Siemens	Mitsubishi
1968	GM Design Spec						
1969		Modicon 084					
1970			PDQ II				
1971			PMC	PC45			
1972		Modicon 184	Prog.Computer				
1973				Logitrol	SYSMAC-M1R		
1974			1774 PLC				
1975		Modicon 284/384					
1976						Simatic S3	
1977		Gould	PLC2		Standard SYSMAC		
1978							
1979		MODBUS	Data Highway			Simatic S5	
1980	Ethernet		PLC3				
1981				Series 6			MELSEC FX
1982	Koyo SR21					Simatic TI305	
1983				Series 1	Host Link		
1984			Rockwell			S5-135U	
1985			PLC5	Series 3			A Series
1986		Modicon 984		GE-Fanuc			
1987				Series 5	C200H		
1988							
1989	Profibus						
1989	TI Series 305						
1990	MS Windows 3.0			GE 90-30			
1991			SLC500	GE 90-70	CV Series		
1992							
1993	Profibus DP		DeviceNet		CQM1		
1994	IEC 61131-3	Quantum Series	Rockwell Software			Simatic S7	
1994	PLC Direct, CodeSys						
1995			MicroLogix 1000				
1996					SYSMAC Link		
1997		Schneider	ControlLogix L1			PCS7	
1998				VersaMax			
1999	Automation Direct				CS1		
2000			ControlLogix L55				
2001			Micrologix 1500		CJ1	Profinet	
2002			ControlLogix L60	Rx7i	CJ1M		Q Series
2003				Rx3i			
2004							
2005			Micrologix 1100		CP1H		
2006		M-340					
2007					CP1L		
2008			Micrologix 1400				
2009						S7-1200, 1500	
2010			ControlLogix L70		CP1E, CJ2M		
2011							
2012							
2013							
2014							iQ-R Series
2015							
2016			ControlLogix L80				

History Bibliography

Babbage's Analytical Engine –

Collier, Bruce (1970). The Little Engines That Could've: The Calculating Machines of Charles Babbage (Ph.D.). Harvard University.

Menabrea, Luigi Federico; Lovelace, Ada (1843). "Sketch of the Analytical Engine invented by Charles Babbage... with notes by the translator. Translated by Ada Lovelace". In Richard Taylor. Scientific Memoirs 3. London: Richard and John E. Taylor. pp. 666–731.

Birth of the Programmable Controller –

Segovia, Vanessa Romero; Theorin, Alfred (2012). History of Control, History of PLC and DCS.

The History of the PLC(as told to Howard Hendricks by Dick Morley) http://www.barn.org/files/historyofplc.html

W. Bolton, Programmable Logic Controllers, Fifth Edition, Newnes, 2009

Early Computer Memory –

Edwin D. Reilly, Milestones in computer science and information technology, Greenwood Press: Westport, CT

Jay W. Forrester, "Digital Information In Three Dimensions Using Magnetic Cores", Journal of Applied Physics 22, 1951

Electromechanical Computers –

Cohen, Bernard (2000). Howard Aiken, Portrait of a computer pioneer. Cambridge, Massachusetts: The MIT Press.

The First Electronic Computers –

Goldstine, H. H.; Goldstine, Adele (Jul 1946), "The Electronic Numerical Integrator and Computer (ENIAC)", Mathematical Tables and Other Aids to Computation

ENIAC specifications from Ballistic Research Laboratories Report No. 971 December 1955, (A Survey of Domestic Electronic Digital Computing Systems)

Redmond, Kent C.; Smith, Thomas M. (1980). Project Whirlwind: The History of a Pioneer Computer. Bedford, MA: Digital Press.

The Evolution of the Personal Computer –

Freiberger, Paul (1982-08-23). <u>"Bill Gates, Microsoft and the IBM Personal Computer"</u>. InfoWorld. p. 22.

Krasnoff, Barbara (3 April 1984). "Putting PC Compatibles To the Test". PC Magazine. pp. 110–144.

Norton, Peter (1986). Inside the IBM PC. Revised and enlarged. New York. Brady.

Ward, Ronnie (November 1983). "Levels of PC Compatibility". BYTE. pp. 248–249.

PLC Improvements –

Webb, John W. (1988), "Programmable Controllers, Principles and Applications", Merrill Publishing Company

Hughes, Thomas A. (2001), "Programmable Controllers", The Instrumentation, Systems and Automation Society

Websites as listed in the Major Platforms section of this book

Relays –

The electromechanical Relay of Joseph Henry
http://history-computer.com/ModernComputer/Basis/relay.html

Physical Layout of a PLC

A block diagram of the physical arrangement of a PLC is shown below:

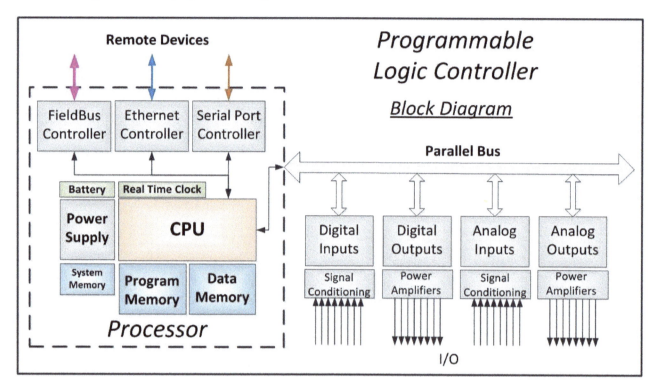

The **CPU** processes all of the logic loaded into the controller and also contains the operating system. It usually has a real-time clock built in, which is used for various functions. The system memory is also associated closely with the CPU.

Not all of the items shown in this diagram are present on every PLC, but this will give you an idea of a typical configuration.

Battery/Memory Back-Up

Program and data memory in a PLC is contained in "RAM" (Random Access Memory). This type of memory may be volatile or non-volatile, and it can (and will) be overwritten often. The program itself is in one area of RAM and must be kept in memory even when the PLC is powered off. Older PLC systems required a battery or a "super-capacitor" to back up the program when power was removed for long periods of time. Newer platforms can save the program to non-volatile memory such as CompactFlash and Secure Digital (SD) RAM. On these older battery-backed platforms, if the battery died, the program was lost.

I/O (Inputs and Outputs)

Physical I/O can be **Discrete**, that is single signals bits that are either on or off, or **Analog**, that is they can be signals that change amplitude in either voltage or current.

Digital/Discrete Devices

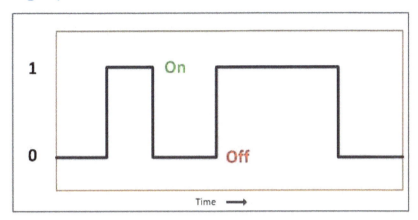

This diagram shows a discrete signal. Typical signal levels for discrete inputs and outputs are 24V DC and 120V AC, but other levels may be present depending on the type of device or input card. In addition to the designation one and zero or on and off, discrete signals may be described as being true or false.

Examples of digital input devices are shown below:

Pushbutton   Photoeye   Proximity Switch

Here are some digital output devices:

Pilot Light   Motor Starter   Solenoid Valve

And here is a device that is both a digital input and a digital output!

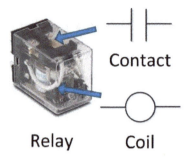

Contact

Relay   Coil

Analog Devices

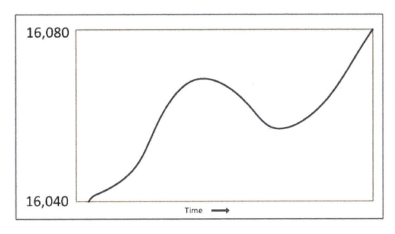

Analog signals vary in either voltage or current. Ranges are typically 0-10V DC or -10 to +10V DC (voltage type) or 0-20mA or 4-20mA (current type). The electrical signal is then converted into a number for use in the PLC program.

Here are some examples of analog input devices:

Potentiometer   Pressure Transducer   Platinum RTD

And these use analog outputs:

Proportional Valve   Linear Actuator   VFD Speed

Analog signals are converted by Analog to Digital Converters (ADCs) for inputs, or Digital to Analog Converters (DACs) for outputs.

ADCs capture a signal from an input device (such as a pressure or temperature sensor) and convert it into a 16-bit Signed Integer. This means that for the full range of the sensor, there are a possible 65,536 different values, ranging from -32,768 to 32,767. Since most sensors only provide positive values, this means that only 0-32,767 are possible. If the ADC has a full 16 bits

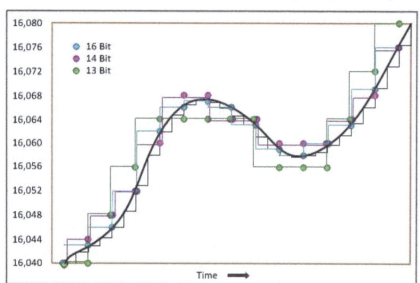

of resolution, this means that every value within that range may occur; however if the converter only has a 14-bit resolution (more common for PLC signals), the values will increment by 4 (i.e. 0-4-8-12 etc.). This means only 8192 possible values can occur! This picture shows the results of converting an analog signal to 16-, 14- and 13-bit resolution.

DACs take a number from the PLC program (memory) and convert it into a signal to control a device such as a proportional valve or a drive (VFD) speed reference. Like ADCs, the signal may be a full 16 bits, but 12-14 bits of resolution are more common for PLC analog output cards.

4-20mA signals are considered to be more immune to noise interference, while 0-10v signals are often used to control VFD (Variable Frequency Drive) speeds.

In the Programmable Logic Controller block diagram, ADCs are shown as "Signal Conditioning", while DACs are shown as "Power Amplifiers".

Wiring - Digital

Digital I/O wiring is based on the type of signal that is connected to the I/O point. Discrete I/O may be AC or DC, Input or Output. Relay cards can also be used to transmit different types of signals.

Discrete solid-state DC devices have two different types, Sinking (NPN), and Sourcing (PNP). NPN and PNP are transistor types. PNP devices *source* a positive DC signal, usually 24v, while NPN devices sink current from a sourcing input card.

These diagrams show typical wiring diagrams for DC devices and sensors.

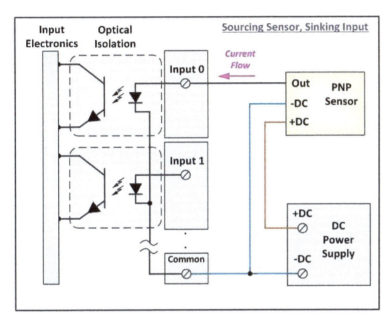

A **Sourcing** (PNP) sensor applies a positive voltage to a Sinking (NPN) input. The input card and the sensor need to share the –DC voltage, which is usually grounded.

The optical isolation protects the lower voltage TTL circuit in the input card. The input card receives its operating voltage (usually 5 volts DC) from the PLC's power supply. The sensor's power is supplied from an external DC power supply, typically at 24 volts DC.

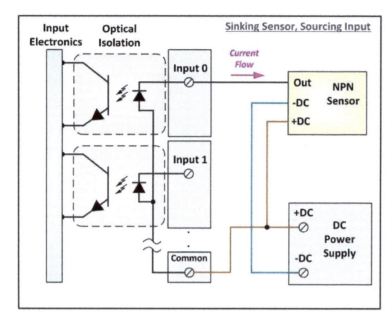

A **Sinking** (NPN) sensor provides a path for current to the sourcing input. As with the sinking input card, it is important that the sensor and input card share the same +DC reference.

DC input cards are often made to allow either PNP or NPN sensors or devices to be used. The card reacts based on the polarity of voltage applied to the common terminal.

Common terminals are also often grouped with inputs, allowing PNP and NPN sensors to be used on the same card if desired. For example, inputs 0-3 may have a positive DC common, while inputs 4-7 may have a negative.

Most U.S. and European machines use PNP sensors and sinking inputs, while NPN sensors are common in Japan and on Japanese equipment.

AC sensors are used with AC type input cards, in this case polarity will not matter. The common terminal will usually be neutral (0 volts, grounded) and the 120 VAC signal will be applied to the input. The main reason AC sensors would be used instead of the safer, lower voltage DC is the relative immunity from noise and the longer distances allowable for wiring. Conveyor systems and large industrial facilities often use AC inputs and outputs.

As with DC inputs, DC outputs may be sinking or sourcing. They are also optically isolated as shown in the diagram.

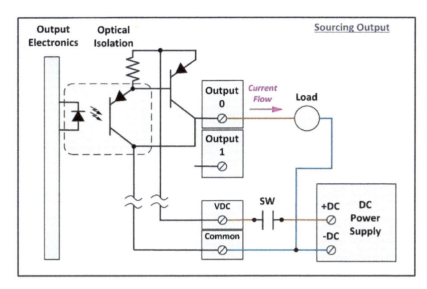

DC output cards require both a positive and negative DC signal to operate. This is because of the power requirements of the outputs; DC outputs are often capable of handling up to 2 amps each.

The VDC connection is often switched as shown by the "SW" contact. This is for safety, which must be hardwired to an Emergency Stop circuit if the load controls a potentially hazardous device. The diagram above shows a sourcing card. The output provides a positive voltage to the load. A sinking card would provide a path for current *from* the device and sink it to the common terminal.

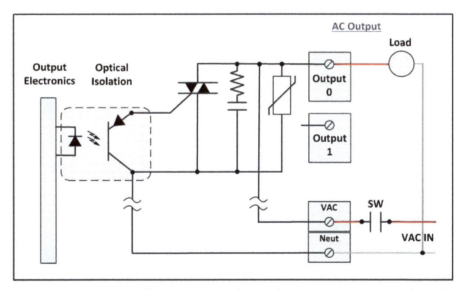

AC output cards usually use a TRIAC to provide AC voltage to the load. Outputs are usually wired through fuses for circuit protection. 120 and 240 VAC cards are available from most manufacturers. They are often used for switching motor starters.

Some AC output cards use zero voltage detection to ensure that the output only switches on when current flow is at zero, reducing surges at the load.

Relay output cards can switch either AC or DC voltages. They often group the outputs with different commons, allowing mixed use cards. The specifications for relay cards show a much lower lifetime than those of solid-state cards because of the mechanical relays.

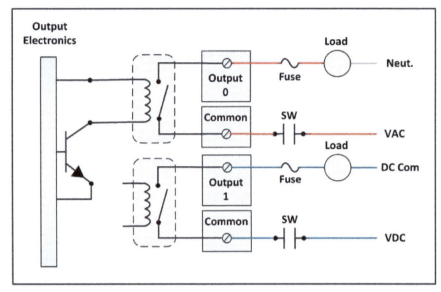

Relay cards that have a common terminal assigned to each output are known as *individually isolated*. This type of card is often used with external equipment that has its own power source.

Since the number of operations of each relay is limited compared to that of a DC output card, external relays are often used instead. Small terminal blockstyle relays are an inexpensive alternative to a relay card.

Wiring – Analog

Analog signals are based on changes in either voltage or current. Voltages may be either 0-5, 0-10, or -10 to +10 volts. Current signals are either 0-20mA or 4-20mA.

Voltage signals are more susceptible to noise, but wiring runs can be longer than with current signals. Analog signals should be connected using shielded cable whenever possible with the shield connected on <u>one end only</u> to reduce noise.

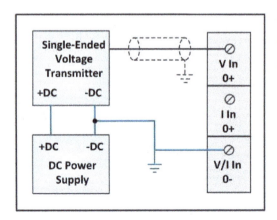

Analog card input configurations may differ on various brands, but this is typical for a universal (voltage and current) input card.

This is a single-ended transmitter, which only has one output wire. It generates the variable voltage signal from the power supply and shares the grounded –DC signal.

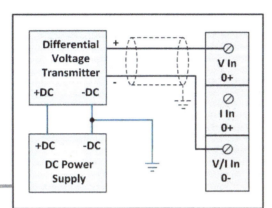

Differential negative two voltage needs to be the negative

transmitters have a positive and a terminal; these are connected to the inputs of the card. While the shield grounded, there is no need to ground lead of the sensor or transmitter.

Two wire current transmitters are placed in-line with the power supply and are also known as *loop-powered* transmitters. They deliver a 4-20 mA signal to the current input of the card.

In this diagram, note the "Anlg Com" terminal; this is a group common termination that is present on many analog input cards. When in doubt, it is always a good idea to jumper the negative side of the power supply to this terminal and ground it.

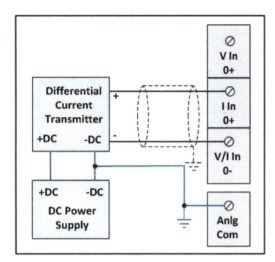

A differential current transmitter is powered by the power supply independently of the current loop. This allows these types of devices to have features like LCD indicators that display current and set-point values. These transmitters are typically more expensive than 2-wire transmitters but have more setup features such as scaling and programmability. The negative lead of the transmitter does not need to be grounded.

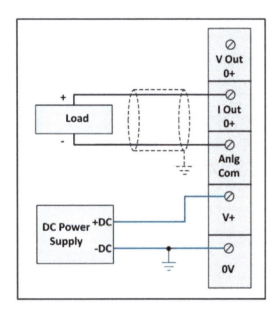

Analog output cards provide either current or voltage to a load. As with input modules, 0-20 and 4-20mA and 0-5, 0-10 and -10 to +10 VDC signals are standard.

This card shows the wiring for a typical analog output to a load. Some cards use an external power supply as shown, while others are powered from the 24v supply on the I/O bus.

As with analog input wiring, it is important to use shielded cabling and ground the shield on <u>one end only</u>.

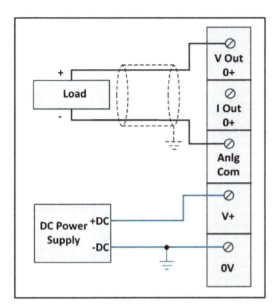

Analog voltage outputs also may be powered with an external power supply or across the bus. An example of a typical load for an analog voltage output would be the speed reference for a Variable Frequency Drive (VFD) or a position for a proportional valve.

Communications

All PLCs need some method of communicating with a programming device, and possibly even with other devices such as Operator Interface Terminals (OITs), computers or remote I/O nodes. A PLC processor will usually have at least one built-in communications port, and possibly more. In addition, modular communication devices can be added to the parallel bus, or rack.

Serial Communications means that bits are transmitted on one wire at a time in a series of high and low electrical signals, or "1's and 0's". This differs from parallel communications, such as printers, where the bits are transmitted and received on many wires in parallel with each other. Serial communications uses different physical formats, usually in the form of RS232, RS422 or RS485.

RS, or "Recommended Standard" communications include RS232, RS422 and RS485 serial communications. RS defines the wiring and format of the message, but not the language or protocol of the message.

RS232:

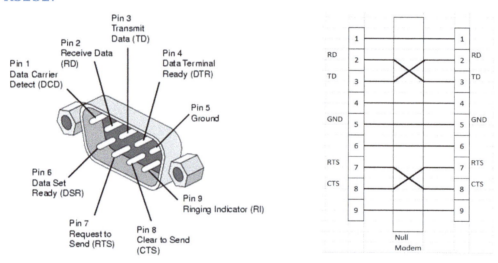

RS232 protocol is used for many of the programming interfaces between laptops and PLCs. Because serial ports are often not present on newer computers, a USB adapter may have to be used between the computer and cable. RS232 requires that parameters such as Baud rate (speed), bits and parity be set the same on all stations for communication to take place. If the transmit and receive pins (TD and RD) are the same on both devices, a null modem adaptor will be needed.

Each PLC manufacturer will have its own driver for its language; for example, in Allen-Bradley's PLC the protocol is called **DF1**, while in Siemens it is called **MPI**.

RS485:

RS485 is another common standard for serial communications. It uses a single twisted pair of wires and can be "daisy-chained" between multiple controllers.

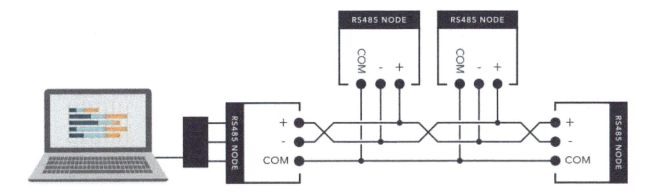

RS485 is commonly used for remote I/O and multi-drop networks. Programming devices such as computers can be used on these networks through the use on an adapter. As with RS232, the Baud Rate and protocol must be set the same on all member stations or "nodes".

Examples of RS485 networks are Profibus, DeviceNet, Modbus and Data Highway/DH+.

RS422:

RS422 is often called peer-to-peer, or PtP. It also uses a twisted pair of wires, but this protocol only supports communication from a single device to the PLC.

A Note on Twisted Pair Communications:

Twisted pairs of wires reduce the amount of electrical interference from other signal wires that run in parallel. Noise can be further reduced by shielding the pairs as a whole or individually; it is important that the shields are only grounded on <u>one end</u>, otherwise even more noise may be introduced.

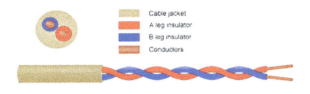

Acronyms are used for the different types of twisted pair cables used for communications.

UTP signifies Unshielded Twisted Pair as shown in the diagram to the left. **STP**, or Shielded Twisted Pair, means that a single shield surrounds all of the pairs in a cable, while **ScTP** or "Screened Twisted Pair" means that individual shields are wrapped around each pair.

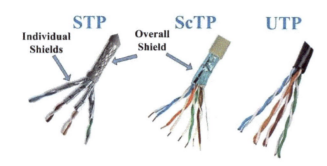

This diagram illustrates some of the types of twisted pair cables used in industrial communications. A bare (uninsulated) wire called a drain is usually run in contact with the foil shield.

USB:

Universal Serial Bus, or USB is a standard developed in the 1990s to connect computer devices and peripherals such as keyboards, digital cameras and pointing devices to computers. It not only provides communications, but also can power devices. USB is much faster than standard serial connections.

Ethernet:

Ethernet is a family of computer networking technologies consisting of a set of wiring and communications standards. Systems communicating over Ethernet divide a stream of data into shorter pieces called **frames**. Each frame contains source and destination addresses, and error-checking data so that damaged frames can be detected and discarded; most often, higher-layer protocols trigger retransmission of lost frames.

Cabling for Ethernet may consist of coaxial cable, several twisted pairs as shown above, or fiber-optics. Most PLC connections use standard twisted pair CAT 5 cable and RJ45 connectors.

Ethernet follows a seven-layer structure defined by the Open Systems Interconnect model. These layers describe the lowest or physical layer, as well as various methods of interconnecting and networking between different areas, or domains.

Many different protocols are included on this definition, including TCP/IP (for connecting dissimilar devices across the internet), BOOTP (for setting initial addresses), and SMTP (for e-mail).

PLC manufacturers generally define their own language and protocols for using Ethernet to both communicate and connect to inputs and outputs (I/O).

Because it is important that I/O communications are **deterministic**, PLC manufacturers follow the Common Industrial Protocol, or CIP. This ensures that signals are sent and received within a specified period of time, and are therefore predictable. Examples of CIP include Ethernet/IP for Allen-Bradley and ProfiNet for Siemens.

Ethernet Terminology:

Following is a list of terms that it is important to know when discussing an Ethernet network:

Client: A computer or device that initiates a request for data.

Server: A computer or device that responds to the client by providing or accepting data.

LAN (Local Area Network): a network that that connects computers or devices in one single location. Usually administered by a Server computer or a domain controller.

WAN (Wide Area Network): a network of LANs connected by Gateway or Router devices.

Workgroup: Computers in a workgroup can share files folders and printers.

Domain: A collection of host computers supervised by a domain controller computer.

Bridge: A device that interfaces between two similar networks.

Gateway: A device that allows modules on two different communication networks to interface.

Hub: A solid-state device that connects Ethernet modules in a star configuration.

Switch: A solid-state device that connects Ethernet modules that has buffering capability to reduce collisions.

Router: A solid-state device that has the capabilities of a switch, but connects Ethernet modules on different networks. Often includes a **Firewall**.

TCP/IP (Transport Control Protocol/Internet Protocol): A common protocol that allows computers with different operating systems to exchange data.

CIP (Common Industrial Protocol): A method of communicating used in industrial automation. Provides deterministic communications for control, safety, synchronization, motion, configuration and information.

Socket: A package of subroutines that provides access to TCP/IP and CIP functions.

NIC (Network Interface Card): A communication adaptor with an Ethernet port in a computer.

Ethernet Addressing:

There are several classes of Ethernet addresses allowing for different numbers of devices to be connected together in the same network. Most industrial control networks use class C, which comprises a single LAN.

The format of an Ethernet address is xxx.xxx.xxx.xxx, where values of xxx can range from 0-255. Each of these sections or "octets" therefore represents a Byte, or 8 bits. An Ethernet address then contains 32 bits.

The Ethernet address is then "Masked" with a group of 1's and 0's to allow only groups with similar numbers to communicate. A class C network will often use numbers in the "192.168.0.xxx" series with a mask of "255.255.255.0".

Industrial Communication and Control:

An example of an Industrial Control Network is shown in the following diagram:

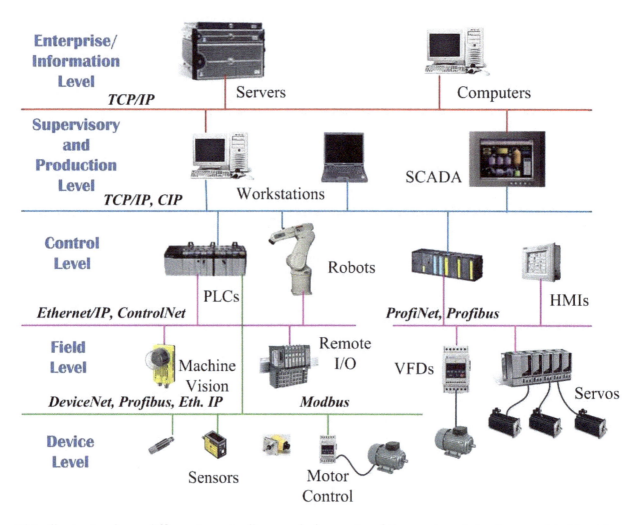

This illustrates how different controllers and elements of the system interface and bridge the various networks. All of the networks below the control level need to respond quickly to changes, while the communications at the higher levels are concerned with transferring and saving large amounts of information.

Exercise 1

1. What kind of PLCs do you have in your plant? Brand name (platform) and type?

   _____
   _____
   _____

2. What type of I/O do they use? (Analog/Digital, voltages, etc.)

   _____
   _____

3. What types of communication networks do you have in your plant?

   _____
   _____
   _____

4. What is the Ethernet address and Subnet Mask of your computer?

   _____

5. On which level of communication network are PLCs usually found?

   _____

PLC Memory

PLC Memory consists of the operating system and firmware of the processor, sometimes called **System Memory**, the module firmware (if any), and the program and data that is used by the programmer. In the previous hardware section it was explained that there are volatile and non-volatile areas of memory, and that the volatile part of memory needs a battery, "super capacitor" or other rechargeable energy storage module to hold its program and/or data.

Though the program can be saved on Flash or SD RAM cards without a battery, the data exchange rate is too slow to use this for the actual interfacing of the program with its data. When the PLC is powered on, the program is loaded from non-volatile RAM cards into the user memory of the controller. Not all PLC platforms back up the user memory with a battery or other energy storage device, data memory may be lost when a processor loses power. Some platforms, however, ensure that the data is kept intact even when power is lost by use of battery backed RAM. This means that the values in data registers will be retained and the program will start in its last state.

Other PLC platforms assign some parts of RAM to be "retentive" and other parts non-retentive. Omron separates its retentive bits into "holding" relays and non-retentive "CIO", and its data into the retentive DM Area and non-retentive "Work Area". Siemens allows its general "marker" memory to be assigned as retentive or non-retentive and defaults to only 16 bytes of retentive marker memory. Siemens data blocks, however, are retentive unless defined not to be. Allen-Bradley's memory is all retentive.

The operating system itself on a processor is held in non-volatile System memory, called "firmware". To change the firmware on a PLC a "Flash" program or tool needs to be used to download it. This is usually included with the programming software.

I/O, communications and other modules also often have firmware built in. The firmware update tools can also update these modules and the firmware is usually available from the manufacturer's website. It is necessary to have software that is at least as up-to-date as the firmware being installed.

The RAM part of memory in a PLC can be separated into two general areas: **Program Memory** and **Data Memory**.

Program memory consists all of the lists of **Instructions and Program Code**. This is what is sent to the processor. The act of sending the program instructions to the PLC is called "downloading" on most brands of PLC, however this may differ on some platforms.

Data memory includes the **Input and Output Image Tables** as well as **Numerical and Boolean Data.** You will find that _most of the data used in the PLC program is internal memory and not directly related to I/O!_

As the program executes, it keeps track of whether bits (BOOLS) are on or off and the values of numbers in Data Memory. Different platforms have different ways of organizing data; some of these methods will be discussed in this section.

Before discussing the way memory is organized, it is important to discuss the types of data that will be stored there.

Numerical Data Types:

BOOL or Bit: This is the simplest form of data, it only has two possible states, on and off, or one and zero. If the states are called "True" and "False", it is more properly called a BOOL, which implies a logical operation. A bit may be just one part of a larger element, such as a byte or word.

Byte or SINT: A Byte is a group of 8 bits. This is sometimes called a Single Integer, or SINT. A Byte's value ranges from 0 to 255, or 0000_0000 to 1111_1111. What is a half of a Byte, or four bits? A **Nibble** of course!

Integer (INT) or WORD: 16 bits make up an integer, or INT. While an Integer always implies a number that mathematical functions can be applied to, a WORD may sometimes only be used for logic functions, such as AND or OR. An Integer can have 65,536 possible values, from 0000_0000_0000_0000 to 1111_1111_1111_1111.

The bit on the far right of the string of 1's and 0's is known as the Least Significant Bit, or LSB, while the bit at the far left is the Most Significant Bit, or MSB. An **Unsigned Integer** ranges in value from 0 to 65,535. A **Signed Integer** uses the MSB as a "sign bit"; if it is a 1 the value is negative, if it is a 0 the value is positive. A Signed Integer ranges in value   from -32,768 to +32,767. Most PLCs use Signed Integers.

Double Integer(DINT) or Double Word(DWORD): A Double Integer has 32 bits. Like an Integer, there are Signed and Unsigned versions, determined by the MSB. A DINT ranges in value from 0 to 4,294,967,295 (Unsigned) or − 2,147,483,648 to +2,147,483,647 (Signed).

REAL or Floating Point: A REAL number is a 32-bit number that can express fractional values, or decimal points. Because the decimal can be shifted to the right or left within the value to change its magnitude, the decimal point can be said to "float". Unlike Bytes, INTs and DINTs, it is not possible to look at individual bit values to determine the value of the number. A REAL number is composed of a Mantissa, an Exponent and a Significand, and the bits within the number are not used for anything else other than expression of the value. REALs have a range of 1.1754944e-38 to 3.40282347e+38.

Additional data types such as Long Integers (LINT, 64 bits), Long Reals, (LREAL, 64 Bits) and STRINGs (an array of Bytes signifying text characters) are also used in PLCs.

Understanding How Bits Can Become Numbers:

If the previous talk of ones and zeros is confusing, think of it this way: for a single bit, there are only two possible values; 0 or 1, off or on.

For two bits, there are four possible values; both off or 0,0; the first off and the second on or 0,1; The first on and the second off or 1,0; and both on or 1,1.

0	0
0	1
1	0
1	1

4's	2's	1's	
0	0	0	0
0	0	1	1
0	1	0	2
0	1	1	3
1	0	0	4
1	0	1	5
1	1	0	6
1	1	1	7

For three bits, the possible number of combinations increases to 8 as shown: the values above the columns show the value of each "place" in the row. The string of ones and zeros is known as binary, a base 2 system. As a new position is added to this string, the value of the previous column is doubled.

For four bits, the possible number of combinations increases to 16, and the value of the next column becomes the "8's" position. The values then start at 0000, or zero decimal, and increase to 1111, or 15 decimal.

The value of the each column then doubles as a placeholder (16, 32…), as does the possible number of combinations (32 @ 0-31, 64 @ 0-63); Remember, computers and PLCs can only "think" or process data in terms of ones and zeros since they are really just a collection of on-off switches, albeit very tiny ones.

Data Formats:

In addition to the data types listed above, data can also be expressed in different ways. For instance, a Byte can be shown as a string of ones and zeros (Binary or Base 2): (0110_1011), as a decimal (Base 10) number: (107) or as a Hexadecimal (Base 16) number: (6B).

Integers: Computers and PLCs operate most efficiently when using numbers that are multiples of two; this is because at its heart, a microprocessor is just a collection of on-off switches. An Integer is therefore just a series of binary values signifying increasing values for each bit as shown below:

(MSB/Sign)															(LSB)
32,768	16,384	8192	4096	2048	1024	512	256	128	64	32	16	8	4	2	1
0	1	1	0	1	0	0	1	0	1	1	0	1	0	0	1

To determine the decimal value of a Signed Integer for the binary value, it is necessary to add up all of the place values where there is a 1: 16,384 + 8,192 + 2,048 + 256 + 64 + 32 + 8 + 1 = **26,985**.

(MSB/Sign)															(LSB)
32,768	16,384	8192	4096	2048	1024	512	256	128	64	32	16	8	4	2	1
1	0	0	1	0	1	0	1	1	0	1	0	1	0	1	0

This Signed Integer value has a 1 in the MSB. The decimal value is determined by adding all of the other place values: 4,096 + 1,024 + 256 + 128 + 32 + 8 + 2 = 5,546; and then subtracting the result from 32,768, or **-27,222**. This is also known as **2's Complement**.

BCD: Prior to the use of OITs (Operator Interface Terminals), digital devices such as thumbwheel switches and 7-segment displays were used for entering and displaying decimal values in the PLC.

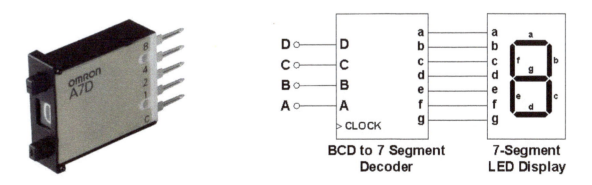

BCD to 7 Segment Decoder   7-Segment LED Display

To enter a number into a PLC's memory location, each thumbwheel switch required 4 digital inputs into the PLC. Decimal numbers from 0-9 could be set for each decimal digit; when the thumbwheel reached 9, it would then roll over to the zero position again. This meant that combinations such as 1010 (10) or 1011 (11) were not possible.

In the same way, outputs were used to illuminate the individual segments of a 7-segment display, each display requiring 4 digital outputs. It was important in this case that illegal combinations of outputs (decimal 10 and above) could not be sent to the device.

Binary Code ABCD	Decimal Number	BCD Code B_5 B_4 B_3 B_2 B_1
0 0 0 0	0	0 0 0 0 0
0 0 0 1	1	0 0 0 0 1
0 0 1 0	2	0 0 0 1 0
0 0 1 1	3	0 0 0 1 1
0 1 0 0	4	0 0 1 0 0
0 1 0 1	5	0 0 1 0 1
0 1 1 0	6	0 0 1 1 0
0 1 1 1	7	0 0 1 1 1
1 0 0 0	8	0 1 0 0 0
1 0 0 1	9	0 1 0 0 1
1 0 1 0	10	1 0 0 0 0
1 0 1 1	11	1 0 0 0 1
1 1 0 0	12	1 0 0 1 0
1 1 0 1	13	1 0 0 1 1
1 1 1 0	14	1 0 1 0 0
1 1 1 1	15	1 0 1 0 1

This coding of bits into decimal equivalents is known as **Binary Coded Decimal**, or **BCD**. Some operator interfaces still require BCD for display of values, so conversion of integers into BCD format is sometimes needed. In addition, there are PLC platforms that use BCD for Timer and Counter values.

As the table to the left shows, BCD codes above 1001 require that another 4 digit value be used for the next decimal number. To display the number 9,999, the sixteen bit pattern would read 1001_1001_1001_1001

The equivalent 9,999 expressed as a Signed Integer would be 0010_0111_0000_1111.

To express a Signed BCD number, the most significant four bits are used as a "sign" character, as with the Signed Integer. A negative number is signified by a 0001, while a positive number uses 0000.

The range of a Signed BCD number is then only -999 to +999!

Binary	Decimal	HEX
0000	0	0
0001	1	1
0010	2	2
0011	3	3
0100	4	4
0101	5	5
0110	6	6
0111	7	7
1000	8	8
1001	9	9
1010	10	A
1011	11	B
1100	12	C
1101	13	D
1110	14	E
1111	15	F

Hexadecimal: In order to display the full 65,536 possible values in sixteen bits in only four characters, a base 16 numbering system called **Hexadecimal** is used. The only reason the base 10 Decimal system is used is that humans calculate best in this format; all because we have 10 fingers and 10 toes. As mentioned before, computers are most efficient when they calculate in multiples of two.

After a group of four binary digits (a "**Nibble**", or half of a Byte) reaches 1001, the next value of 1010 can't be expressed as a numeral without using something outside of the values 0-9. In order to describe a base 16 number using 16 different symbols, after the number 9 the symbols use the letters A-F as shown to the left.

Because Hexadecimal is base 16 and a multiple of 2, it is very easy to convert from Binary to Hexadecimal. Simply separate the binary into groups of four and convert each group.

Octal: Base 8 numbers are also commonly seen in PLC systems. For instance, Siemens I/O addressing is octal; this means that only the numbers 0-7 are used. The numbers after 7 would be 10, 11 …17, 20, 21 and so on.

Like Hexadecimal, because Octal is a multiple of 2 it is very easy to convert to and from Binary. Separate the binary into groups of three and convert each group; the highest value in any group of three is 111, or 7.

Viewing Data Types: PLC software allows the user to view a number in many of the formats listed here, removing the need for a calculator. It is not always clear in what format the data is being viewed, but there are often designators. In Siemens, a signed integer (decimal or Base 10) will have no designator, however a Hex number will have a W#16 prefix, indicating that it is base 16. A REAL will have a decimal point or be expressed with an exponent, while a binary representation may have a prefix or appear as a string of ones and zeros.

Dot Fields and Separators: If a single bit of an integer is designated, it may be shown with a separator such as a slash or a dot; for example N7:5/3 (Allen-Bradley, the fourth bit of the sixth word; numbering starts at zero) or Q3.2 (Siemens, the third bit of the fourth output byte).

Dot fields are also often used to designate an element of a complex data type, such as a timer. For example Timer1.ACC designates the accumulated value (integer or double integer) of Timer 1. It is important to understand how memory is addressed for your particular PLC before beginning a program.

Tags: Many modern PLC platforms don't use numerical data registers at all. Instead, they allow users to create memory objects as required in the form of text strings. Allen-Bradley's ControlLogix and Siemens' TIA Portal platforms are examples of this. Most major PLC manufacturers make a PLC with tag-based data. Tags are also called Symbols on some platforms, but a symbol is not necessarily a tag; it may simply be a mnemonic address or shortcut to a register address. Tagnames are downloaded into the PLC and used instead of an address.

Tags are usually created in a data table as required. Instead of numeric addresses such as "B3:6/4" or "DB2.DBW14", symbolic names such as "InfeedConv_Start_PB" or "Drive1402.ActualSpeed" are created as memory locations. As tags are created, details such as the data type (BOOL, Timer, REAL) and the display style (Hex, Decimal) need to be selected.

Tags have the advantage of being more descriptive than numerical register numbers. In addition, descriptions and symbols from register addresses were only present in the computer and were not downloaded into the PLC. With tags, since the address is the actual register location, a tag-based program can usually be uploaded straight from the PLC.

Also, the same tags from the PLC program can be used in the HMI or SCADA program directly. This saves time rather than having to map PLC addresses to HMI tags.

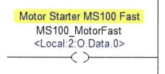

Of course, I/O addresses are still created from the hardware configuration of the PLC, but manufacturers have created various ways to connect I/O addresses with tags. One of the most useful of these is Allen-Bradley's ControlLogix platform, where any tag or address can be "**aliased**" to any other and both show up in the ladder logic, as shown in the figure to the left. Modicon's newer platforms also allow "Symbols" or tags to be connected to I/O points in a similar way.

Data Structures:

Array: An Array is a group of <u>similar</u> data types. For instance, an array can be defined that contains ten integers, or 50 REALs, or 32 BOOLs. Data types can't be mixed in an array.

Complex data types such as Timers, Counters or User Defined Types (UDTs) can also be placed into an array. Typically an array will be shown with square brackets, such as Delay_Tmr[6]. This designates element number 6 of the array.

Some platforms allow multidimensional arrays to be defined, such as Integer [2,4,5]. This means Integer number five of the fourth group of the second group.

Elements that are composed of more than one data type are known as **Structures**. A structure may be defined by the programming software, as with instructions, or by the programmer.

User Defined Type (UDT): A UDT is a group of <u>different types</u> of data, or a structure. Later in this course, data types that consist of more than one type of data will be discussed; for instance Timers and Counters consist of two integers or double integers and several bits, all combined into a structured data type called "Timer" or "Counter".

A UDT can only be used with Symbols or Tags; this is because a UDT is not data. Instead, it is a *definition* of data. After defining a UDT, a tag or symbol must be created using the new data type.

A common reason to build a UDT is to describe an object more complex than a simple element of data. For example, a Variable Frequency Drive has many pieces of data that may be associated with it. For instance, a motor needs to start and stop. It has various numerical parameters to describe its movement such as commanded speed, actual speed, acceleration and deceleration. We may also want to know its status; whether it has faulted and what kind of fault has occurred.

UDT Name:	*"Drive"*			Name	DataType	Status	Description
Name	**Type**	**Description**		Drive_5207	"Drive"		Spindle Drive VFD 5207
Run	BOOL	Run Command		Drive_5207.Run	BOOL	0	Run Command
Stop	BOOL	Stop Command		Drive_5207.Stop	BOOL	1	Stop Command
Alarm	BOOL	Drive in Alarm		Drive_5207.Alarm	BOOL	1	Drive in Alarm
Running	BOOL	Run Status		Drive_5207.Running	BOOL	0	Run Status
CMD_Speed	REAL	Commanded Speed %		Drive_5207.CMD_Speed	REAL	27.34	Commanded Speed %
Act_Speed	REAL	Actual Speed %		Drive_5207.Act_Speed	REAL	0.00	Actual Speed %
Accel	INT	Acceleration ms		Drive_5207.Accel	INT	40	Acceleration ms
Decel	INT	Deceleration ms		Drive_5207.Decel	INT	50	Deceleration ms
AlarmStatus	SINT	Alarm Number		Drive_5207.AlarmStatus	SINT	4	Alarm Number

On the left is a UDT named "Drive" defined in the software, and on the right is a tag made from the UDT. The definition does not get downloaded to the processor, it can only be modified on the programming device.

The sub-elements of the tag are an example of the dot fields described earlier. By making a UDT, many drives can be added to a program without a lot of extra typing. UDTs are in important element in Rapid Code Development.

Tip: On non-tag-based systems, UDTs can cause problems if the commented program is not available. Remember, descriptions and non-tag symbols are not kept in the processor. This is why it is difficult to reconstruct a Siemens S7 program if you don't have the original code.

Exercise 2

1. Convert the Binary number "0110_1100_1011_0111" into...
   Decimal _____   Hexadecimal _____   Octal _____

   Can this number be converted to BCD? Why or why not?

   _____

   _____

2. How would you write the Binary number "1001_0101_1000_1001" as a Signed Integer?

   _____

3. Convert the Decimal number 417 into:
   BCD _____   Binary _____   Hexadecimal _____

4. Write the number 2A9E in Binary : _____
   What is this number in Decimal? _____

5. How many Bytes are in a Double Integer? _____

6. What is a "Tag"? Is it the same as a Symbol? _____

   _____

Work Area:

Data Memory Organization:

Data memory is organized in different ways depending on the type or brand of PLC. Some PLCs have registers assigned to specific data types, that is, Bit, Integer or Real, (Allen-Bradley SLC and Micro), whereas other brands may separate data by whether it is retentive (Holding Relays, Omron) or place all data together ("V Memory", Koyo/Automation Direct).

It is important when learning a new PLC's programming platform to first understand how its memory is organized. In the older GE PLCs for instance, data memory and I/O share the same space. It could be quite embarrassing if you were to cause actuators to move when intending to simply save an integer to a register!

A-B SLC		Siemens S7		Omron		Koyo		
O	Outputs	I	Digital In	CIO	Basic I/O (Discrete)	T	Timer Curr. Val	V0-377
I	Inputs	Q	Digital Out	CIO	Special I/O (Analog)		Data Words	V400-777
S	System	PIW	Analog In	CIO	CPU Bus I/O	CT	Counter Curr. Val	V1000-1377
B	Bits	PQW	Analog Out	W	Work Area		Data Words	V1400-7377
T	Timers	M	Memory (M)	H	Holding Area		System	V7400-7777
C	Counters	DB	Memory Blocks	A	Aux Relay Area		Data Words	V10000-35777
R	Control			TR	Temp. Relay Area		System	V36000-37777
N	Integers			D	Data Memory Area	GX	Remote Inputs	V40000-40177
F	REALs			E	Ext. Memory Area	GY	Remote Outputs	V40200-40377
				T	Timers	X	Disc. Inputs	V40400-40477
				C	Counters	Y	Disc. Outputs	V40500-40577
				TK	Task Flags	C	Ctrl. Relays	V40600-40777
				IR	Index Registers	S	Stages	V41000-41077
				DR	Data Registers	T	Timer Status	V41100-41117
						CT	Counter Status	V41140-41157
						SP	Special Relays	V41200-V41237

This table shows the layout of several PLCs' memory areas. The first list, **Allen-Bradley's SLC** and **Micrologix** family, shows that data is segregated into numbered files, O0, I1, S2...F8. Each data file is expandable up to 255 words, but after that, new file numbers have to be added, for instance N9, B10 and so on.

The next table shows **Siemens' Step 7**. I/O is assigned during hardware configuration rather than by slot number as with Allen-Bradley. The general memory area "M" is of a fixed size, whereas Memory Blocks or Data Blocks (DBs) contain a mix of data types and can be up to 64KB in size!

Omron's memory sizes are fixed in dimension for each data type, memory is not allocated dynamically as in the previous two examples. It is unique in that it separates retentive memory (Holding Area) from non-retentive (Work Area).

Koyo (Automation Direct) uses a large data area much as the GE system described earlier, each type of data is fixed in size and can't be expanded. All data can be accessed by direct addressing, the "V" addresses.

I/O Addressing:

I/O addressing varies from brand to brand. Inputs may be addressed as I or X, outputs as O, Q or Y, and analog I/O designations may use a completely different format than digital ones.

Some brands, such as Allen-Bradley, designate I/O based on the slot number where the card is assigned while configuring hardware; this can't be changed. Other platforms, like Siemens, have a default location where I/O is assigned during configuration, but this can be overridden by the programmer. Addressing may also be Octal, Decimal or even Hexadecimal!

	Allen-Bradley SLC-500	Allen-Bradley ControlLogix	Siemens S7	GE 311	Omron CP1E	Mitsubishi FX2N	Codesys
Digital Input	I:1/3	Local:1:I.Data.3	I0.3	I0103	I0.03	X003	%IX4000.3
Digital Output	I:2/5	Local:2:O.Data.5	Q3.5	Q0205	Q1.05	Y025	%QX4002.5
Analog Input	I:3.2	Local:3:I.Ch1Data	PIW272	AI06	CIO200	D302	%IW2022
Analog Output	O:4.3	Local:4:O.Ch2Data	PQW800	AQ007	CIO210	D403	%QW2036
Address Base	10	10	8	10	10	8	

The table above shows some examples of I/O addressing from several platforms.

Program Memory Organization:

All PLC platforms have a routine that is designated to run first. It is important when learning a new platform to identify which routine this is. This is sometimes known as the "Main Routine".

As with data memory, the program itself can be organized in different ways. Major PLC platforms all have some form of subroutine, though they may be called by different names. Allen-Bradley's PLC5 and SLC500 platforms organize their subroutines by file number, where file 2 or "Ladder 2" is the routine that runs first and calls the other routines. As new routines are created, they are called Ladder 3, Ladder 4, etc.

Siemens' routines can take different forms; OBs or Organization Blocks each have a special purpose based on their number. For instance, OB1 is the continuous routine that runs first, OB86 runs if there is a network fault, and OB35 runs on a periodic basis set by the programmer, such as every 100 milliseconds. Functions, or FCs, are much like standard subroutines, however they have local temporary memory. Function Blocks, or FBs, are like FCs but have retentive memory in the form of the data blocks mentioned earlier.

Koyo or Automation Direct's PLCs have subroutines, but they also have special routines called "stages" that automatically deactivate the stage from which it was called. Omron has subroutines called "sections" and also periodic routines called "tasks".

Tip: It is very important to consider whether memory will be **Global** (available to all programs and routines) or **Local** (only available to part of the program) before beginning a program. Think about whether you will have multiple instances of the same code.

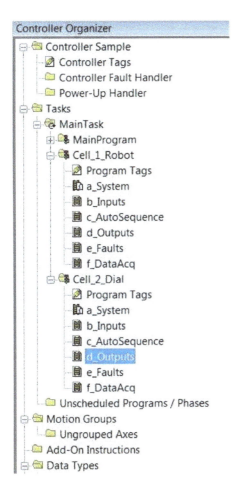

Larger PLCs may also allow multiple programs to be placed into a task, as shown in this Allen-Bradley ControlLogix Controller Organizer figure. While the programs are still scanned one at a time, this allows data tables or tag lists to be assigned to one program, rather than being global. Programs are then scheduled to run in a specific order under the task.

This also allows programs to be duplicated under different names, but with the same tag names. This allows for rapid code development, since a program can be written and tested, then copied, addresses and all.

Tasks can also be assigned to run on a periodic basis as with the OB35 block mentioned earlier in the Siemens description. Analog I/O processing is often done this way to accommodate PID instruction cycle frequency.

Hardware Configuration

PLCs are manufactured in a wide variety of configurations. The simplest "brick" types may only have a few digital inputs and outputs, with no analog and no expansion capability. Sometimes these are called "Smart Relays". The picture to the left shows a small "brick" PLC from Panasonic.

Other PLCs have a very large footprint and are placed into a rack with many slots for expansion. Remote racks can be added to extend the parallel bus and may allow for thousands of local I/O points. Communication modules can further extend the I/O capability as shown in the network diagram in the Communications section.

The picture to the left shows an Allen-Bradley PLC-5 rack, while the one on the right is a Siemens S7-400. Both of these allow for multiple (additional) rack configurations.

In addition to digital and analog I/O, specialty modules for motion control, high speed counting, PID or process control and various other purposes can be added to the rack. Multiple processors or communications modules can also be configured.

There are also many configurations that fall between these very small and very large systems. Some of the most common PLCs are mid-range rack or extendible bus versions. The PLC at the left is a Koyo (Automation Direct) DL405.

Processors for these mid-range modular PLCs may have I/O built in or be stand-alone; large rack systems' CPUs do not have I/O built in. Also, rather than having a physical rack PLC modules may connect together end to end. While this means modules in the middle can't be easily removed, these types of PLC are typically less expensive, and spare slots don't have to be allocated for expansion.

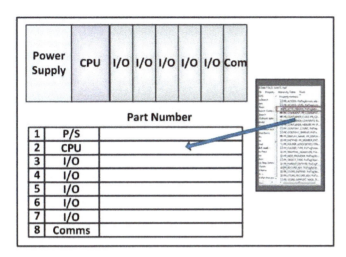

The first step in beginning a new PLC program is configuring the hardware. This is because different processors have different amounts of memory, and because the addresses for I/O are determined by the configuration. As cards are added, new addresses or tags are generated and available for selection in the program.

You can't write a program until you know the I/O addresses and memory configuration!

Some platforms assign I/O addresses by location in the rack (Slot Number) while others allow the programmer to assign addresses. Usually there will be a default address that can be modified in the properties of the card. As explained later in this course, some brand's memory allocation can overlap the I/O area, so careful configuration and planning is important. Selecting hardware will often also include entering a hardware or firmware number for each card. If a rack is used, selecting the rack size will be necessary before inserting cards.

After selecting the hardware for a PLC, the CPU, I/O Cards and communication ports will have additional configuration that will need to be done. Assigning memory locations for status and clock bits, scaling analog channels, and determining failure states of I/O are examples of this. A common method of additional configuration is right clicking the element and selecting "Properties", or simply double-clicking on the card.

Slot		Module	Order number	Firmware	MPI address	I address	Q address
1		PS 307 10A	6ES7 307-1KA00-0AA0				
2		CPU 313C-2 DP	6ES7 313-6CE01-0AB0	V2.0	10		
X2		DP				1023*	
2.2		DI16/DO16				124...125	124...125
2.4		Count				768...783	768...783
3							
4		DI8/DO8x24V/0.5A	6ES7 323-1BH00-0AA0			0	0
5		AI2x12Bit	6ES7 331-7KB02-0AB0			272...275	
6		CP 343-1 Lean	6GK7 343-1CX10-0XE0	V1.0	11	288...303	288...303
7							
8							

This is an example of a Siemens Step7 configuration screen in the Hardware Configuration. Notice the I and Q addresses, these can be changed by double-clicking the card and de-selecting the "default" box. Cards are selected from a catalog with folders for each type of device.

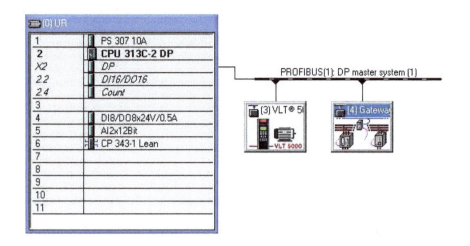

This is from the same Step7 Hardware Configuration screen; notice that the Profibus I/O network can also be configured from here.

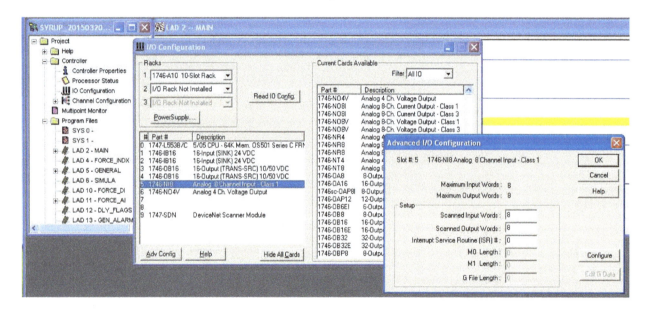

This is an example of Hardware Configuration from Allen-Bradley's RSLogix 500, which is used to program the SLC500 series and Micrologix. Double-clicking the analog card brings up the advanced configuration for the card as shown. Unlike the Siemens configuration, the addresses can't be changed, they are assigned by slot number.

Also, configuring the DeviceNet card in slot 9 requires a separate programming software called RSNetworx. Because of this the network does not show here as it does in the Step7 software.

Exercise 3

1. List 3 different numbering systems that PLC manufacturers may use for addressing data registers:

_____

_____

_____

2. Where does a PLC program begin its scan? (In which routine?) _____

_____

3. Before writing code in a PLC, the _____ must be configured.

4. Besides selecting cards in a PLC, what other kinds of things can be configured?

_____

_____

_____

_____

Program Processing

IEC 61131

PLCs have evolved in different ways depending on the manufacturer. Programming software and methods of handling data can differ immensely from platform to platform. Because of this, in 1982 the International Electrotechnical Commission (IEC) created an open standard that defines what equipment, software, communications, safety and other aspects of programmable controllers should look like. After the national committees had reviewed the first draft, they decided it was too complex to treat as a single document. They originally split it into five sections as follows:

Part 1 - General Information

Part 2 – Equipment and Testing Requirements

Part 3 – Programming Languages

Part 4 – User Guidelines

Part 5 - Communications

Currently, the standard is divided into 9 different parts, and a tenth is being worked on.

The third part, **IEC 61131-3**, defines the languages that are used in programming. It describes two graphical languages and two text languages, along with another graphical method of organizing programs for sequential or parallel processing. It also describes many of the data types described previously in this document.

The first two graphical languages described are Ladder Logic (LAD) and Function Block Diagram (FBD). The text languages are Structured Text (ST) and Instruction List (IL), while the organizational method described above is Sequential Function Charts (SFC), which is also graphical. An additional extension language is Continuous Function Charts, (CFC), which allows graphic elements to be positioned freely; it can be considered as an extension of SFC.

The following examples illustrate the five IEC programming languages; the addresses used are generic and the logic shows selection of Auto and Manual modes, along with a timer enabling "Cycle". These examples do not come from an actual programming language or brand, but are meant to illustrate uses of the languages.

The following examples perform the same function written in all 5 IEC languages.

Ladder Logic (LAD):

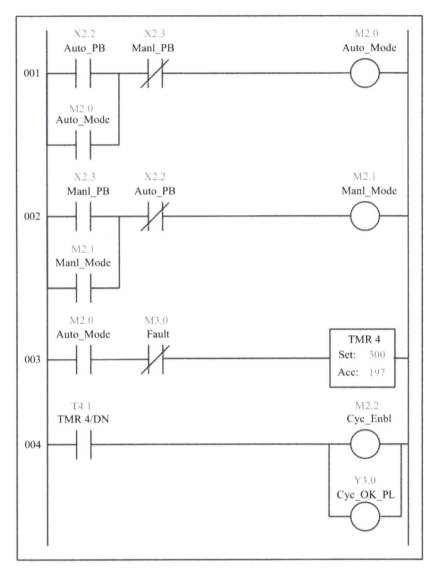

Ladder Logic evolved from electrical circuit drawings, which resemble the shape of a ladder when drawn. As a graphical language, the instructions represent electrical contacts and coils; the vertical sides of the ladder diagram are known as "rails" and the horizontal circuits are often called "rungs". In Siemens software, the rungs are known as "networks".

The "X" addresses represent physical inputs, while the "Y" address is a physical output. The "M"s are internal memory bits.

Because of the variety of addressing schemes as described previously, register representations can mean different things on different platforms.

When monitoring Ladder Logic in real time, usually contacts and coils change color to indicate their state in the logic. If a path of continuity exists from the left rail to the coil, the address will be said to be "On" or "True".

The timer shown in the diagram may also show a time base. If the above timer's preset is three seconds, the time base would then be 10 milliseconds.

More on ladder will be discussed in a later section.

Function Block Diagram (FBD):

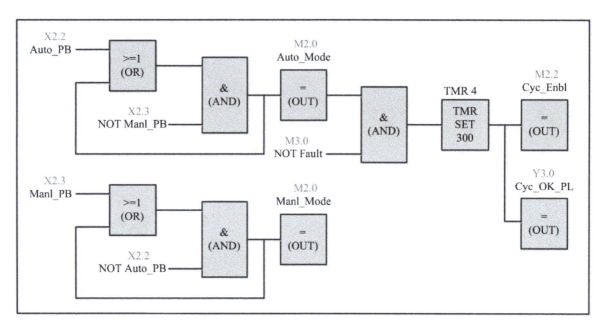

The diagram above shows the same functionality as that in the Ladder Logic diagram. The function blocks evolved from Boolean algebra, the AND and OR representing basic logic. More complex blocks are used for math, loading, comparing and transferring data, timing and counting. As with the previous example, this does not represent any particular brand of PLC.

There are some functions, such as XOR (Exclusive OR) that cannot be easily represented in Ladder Logic. Also, because of the complex nature of some FBD drawings, logic can often extend across many pages. Off-page connector symbols are used to show these connections.

Instruction List (IL):

```
LD X2.2 Auto_PB
O M2.0 Auto_Mode
AN X2.3 Manl_PB
= M2.0 Auto_Mode
LD X2.3 Manl_PB
O M2.1 Manl_Mode
AN X2.2 Auto_PB
= M2.1 Manl_Mode
LD M2.0 Auto_Mode
AN M3.0 Fault
= TMR 4 Set 300
LD T4.1 TMR 4/DN
= M2.2 Cyc_Enbl
= Y3.0 Cyc_OK_PL
```

Graphical languages are usually converted into a text language called Instruction List before being compiled into another low-level code called Machine Language. Before the advent of personal computers, handheld programmers were used to type instructions into the PLC before compilation. These devices often had pictures of Ladder Logic contacts on the keys.

Some platforms, such as Siemens, make extensive use of IL in programming; Siemens version of Instruction List is called Statement List, or STL. Statement List has many commands that are not possible in Ladder Logic or FBD. Other platforms simply use Instruction List as a "stepping stone" to machine language; Allen-Bradley is one of these. Because everything can be converted to Instruction List, yet the opposite case is not true, it is considered to be more efficient than FBD or Ladder.

The table below shows an example of **Assembly Language**, a close relation of machine code. Note the list of addresses and Hexadecimal equivalents to the assembly instructions; this is very similar to the conversion of IL to machine language.

Address	Assembly	Hex	Comment
6050	SEI	78	Set Interrupt Disable Bit
6051	LDA #$80	A9 80	Load Accumulator HEX 80 (128 Decimal)
6053	STA $0315	8D 15 03	Store Accumulator to Address 03 15
6056	LDA #$2D	A92D	Load Accumulator HEX 2D (45 Decimal)
6058	STA $0314	8D 14 03	Store Accumulator to Address 03 14
605B	CLI	58	Clear Interrupt Disable Bit
605C	RTS	60	Return from Subroutine
605D	INC $D020	EE 20 D0	Increment Memory Address D0 20
6060	JMP $EA31	4C 31 EA	Jump to Memory Address EA 31

Tip: Since IL is text based, it is easy to manipulate in third party text or spreadsheet editors such as Microsoft Excel. Instruction List can usually be imported or exported to and from PLC software in the form of .csv (comma-delimited) files or XML (eXtensible Markup Language). This makes it easy to create tables of addresses or tags with a common structure and then convert it into many repetitive rungs or blocks with different addresses. When writing large amounts of repetitive code, this can be a big time-saver!

Structured Text (ST):

```
// PLC Configuration

CONFIGURATION DefaultCfg

VAR_GLOBAL
        Auto_PB      :IN @ %X2.2      // Auto Pushbutton
        Manl_PB      :IN @ %X2.3      // Manual Pushbutton
        Cyc_OK_PL    :OUT @ %Y3.0     // Cycle OK Pilot Light
        Auto_Mode    :BOOL @ M2.0     // Automatic Mode
        Manl_Mode    :BOOL @ M2.1     // Manual Mode
        Cyc_Enbl     :BOOL @ M2.2     // Cycle Enable
        Fault        :BOOL @ M3.0     // Machine Fault
        TMR 4        :TIMER @ T4       // 10ms Base Timer
END_VAR

END_CONFIGURATION

PROGRAM Main

STRT  IF (Auto_PB=1 OR Auto_Mode=1) AND Manl_PB=0 THEN Auto_Mode=1
      ELSE IF (Manl_PB=1 OR Manl_Mode=1) AND Auto_PB=0 THEN Manl_Mode=1
      End IF

      IF Auto_Mode=1 AND Fault=0  THEN
      START TMR 4
      END IF

      IF TMR 4.ACC GEQ 300 THEN
      Cyc_Enbl=1
      Cyc_OK_PL=1
      END IF

      JMP STRT

END_PROGRAM
```

Structured Text resembles high-level programming languages such as Pascal or C. Variables are declared as to data type at the beginning of routines as well as configuration of other parameters. Comments are shown in this program as starting with "//"; this may differ depending on the brand.

Linear programming languages such as Structured Text use constructs like "If-Then-Else", "Do-While" and "Jump" to control program flow. In these languages, syntax is very important and it can be difficult to find errors in programming. Debug tools allowing for partial execution of the code one section at a time are common.

While writing PLC code in Structured Text can be difficult, it is also a much more powerful language than Ladder Logic or Function Block. Libraries can be developed to perform complex tasks such as searching for data using SQL or building complicated mathematical algorithms. At the same time, since the program proceeds step by step, it is more difficult to respond to multiple inputs at the same time; program control can be complex with many loops.

Sequential Function Charts (SFC):

SFC makes use of blocks containing code that typically activates outputs or performs specific functions. In many platforms, the blocks or "steps" can contain code written in other IEC programming languages such as Ladder or FBD. The program moves from block to block by means of "transitions", which often take the form of inputs.

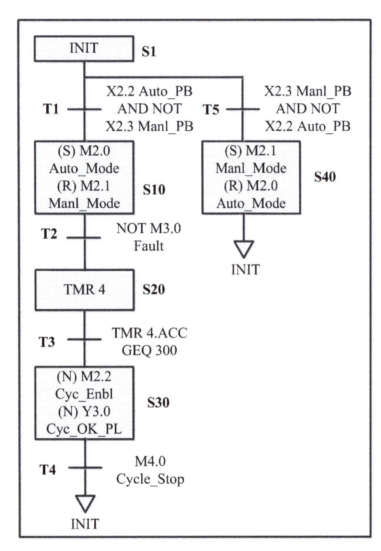

SFC is based on Grafcet, a model for sequential control developed by researchers in France in 1975. Much of Grafcet is, in turn, based on binary Petri Nets, also called Place/Transition nets. Petri Nets were developed in 1939 to describe chemical processes.

Steps in an SFC diagram can be active or inactive, and actions are only executed for active steps. Steps can be active for one of two reasons; either it is defined as an initial step, or in was activated during a scan cycle and not deactivated since. When a transition is activated, it activates the step(s) immediately after it and deactivates the preceding step.

Actions associated with steps can be of various types, the most common being Set (S), Reset (R) and Continuous (N). N actions are active for as long as the step is, while Set and Reset operate as in the other PLC languages.

Actions within the steps and the logic transitions between them can be written in other PLC languages. Structured Text is common in the action blocks, while ladder is often used for transitions. Steps and transitions are labeled as S# and T#. The top of the program will always contain an initial step; the program starts here and this is also where it returns after completion. A program will scan the logic in a step continuously until its associated transition logic becomes true; after this the step is deactivated and the next step is activated.

Scanning

The PLC processor or CPU controls the operating cycle of the program. The operating cycle, or **Scan**, consists of a sequence of operations performed sequentially and continuously.

There are four parts to the scan cycle as listed below:

1. Read physical inputs to the Input Image Table.

2. Scan the logic sequentially, reading from and writing to the memory and I/O tables.

3. Write the resulting Output Image Table to the physical outputs.

4. Perform various "housekeeping" functions such as checking the system for faults, servicing communications and updating internal timer and counter values.

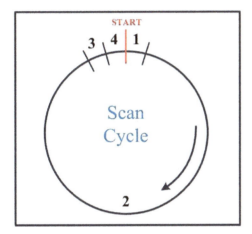

The scan cycle can be visualized as shown in the diagram on the left. When the processor is placed into "run" mode, the state of the analog and digital inputs is captured and saved into memory registers dedicated to the configured inputs. In the second part of the scan, the logic is evaluated sequentially. If the program is written in ladder logic, the rungs are evaluated one rung at a time, left rail to right, top to bottom. As the logic is processed, output and memory coils are energized or de-energized and their status is saved into their respective registers. In the case of the outputs, they are saved into the Output Image Table, which is generated during hardware configuration.

The time that it takes to execute a scan depends on the number and types of instructions in the program. Scans may be as short as 3-5 milliseconds (a very short program or a very fast processor) or as long as 60-70 milliseconds (a longer program). If the scan time exceeds this period by very much, physical reactions of actuators start to become noticeable; it may be time to evaluate a change in PLCs or multiple processors.

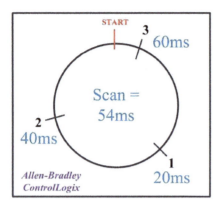

An exception to the scanning method described previously is that of Allen-Bradley's ControlLogix platform. Instead of accessing the Input and Output Image Tables at the beginning and end of the scan, each I/O card is configured with a "Requested Packet Interval", or RPI. I/O tables are updated at this rate, as shown at the left.

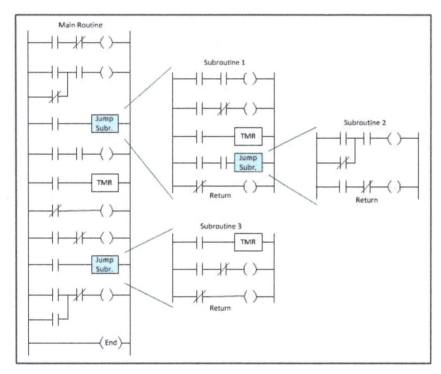

Scanning begins with the first rung of the routine designated as the main routine; all PLCs have a routine with this designation. As logic is processed, calls to other routines will occur as shown. After a subroutine is scanned, the scan returns to where the jump or call was made. Eventually, the scan always ends at the end of the main routine.

In the diagram at the left, the scan begins at the top of the main routine and is evaluated left to right, branch by branch, top to bottom. When the scan reaches a Jump Subroutine (also known as a "call"), the scan continues in the new routine, in this case Subroutine 1. Subroutine 1 then calls Subroutine 2, when the end of Subroutine 2 is reached, the scan continues in Subroutine 1. When the end of Subroutine 1 is reached, the scan continues back in the main routine.

Subroutine 3 is called, is scanned to the end and returns to the Main Routine. When the end of the Main Routine is reached, the resulting Output Image Table is written to the physical outputs and the housekeeping functions (Step 4) is completed. The scan then starts over at the top of the Main Routine.

Scan times will vary from cycle to cycle, depending on the number of instructions and routines active at any given time, the load on the processor from "housekeeping" activities, and communications' connections. Execution times for each type of instruction can be found in each manufacturer's documentation.

Hardware also affects scan times; newer models of processor are much faster than older ones due to advances in technology. For example, Allen-Bradley's new ControlLogix processors are nearly 150 times as fast as their older PLC5, which was the most powerful platform they made in the 1990s!

PLC Modes

PLCs can be placed into a number of operational states. When the processor is executing its program and scanning in the normal way it is said to be in "Run" mode. When initially downloading a program, the CPU is in "Stop" or Program mode, it is not executing the program and I/O is not changing state.

The mode of a PLC can often be changed by means of a switch on the front of the processor, for security this may be a key. There is often a third position on the keyswitch labeled "Remote"; this allows a computer to be used to change the state or mode. This adds an additional layer of safety; when the switch is in Program or Run the state cannot be changed from the computer.

While the computer is connected, the program can sometimes be changed while the PLC is scanning. Not every PLC allows this, and it is not always done in the same way. Some platforms such as Allen-Bradley allow each line or rung of code to be changed while online followed by accepting, testing and assembling (compiling) the program, while others such as Siemens allow each block to be compiled and downloaded without interrupting program execution.

Additionally, many platforms allow the processor to be placed in "test" or "debug" mode. This allows breakpoints to be placed in the code to stop execution at that point. This can be useful if monitoring "looped" code.

I/O can also be "forced" on many platforms. While this is not a mode as such, it does change the operation of the program.

Exercise 4

1. List the five languages defined in IEC 61131-3

_____

_____

_____

_____

_____

2. List the four parts of the scan cycle in a PLC

_____

_____

_____

_____

3. Can a PLC program be changed while it is running? _____

Ladder Logic

The most common and first PLC programming language is Ladder. This is because it is based on physical wiring diagrams, which most electrical maintenance personnel are familiar with. This makes this graphical type of programming easy to read and troubleshoot.

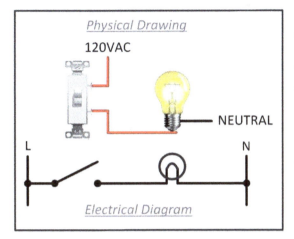

The simplest circuit to understand in an electrical system is a switch turning on a light. The top picture shows a physical drawing of a light switch passing 120 Volts AC to an incandescent light bulb; the bulb also needs a neutral wire to complete the circuit.

The diagram on the bottom shows the electrical equivalent of the physical drawing. The L symbolizes **Line** voltage and the N signifies **Neutral**.

There are many symbols used in schematic diagrams to signify the type of device used in an electrical circuit. The device on the left is generally known as a **Switching Device**, while the device on the right is called the **Load**.

Discrete Logic:

Just like electrical circuit drawings, there are symbols or figures signifying the type of device used in the circuit, however they are classified by their function rather than by the type of device; i.e a pushbutton, switch or sensor uses the same contact icon.

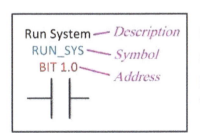

For the logic diagrams in this section, the conventions to the left are used. The **Address** is the register that is updated as the logic is processed. This applies to non-tag-based systems, however even they use addresses for I/O.

The **Symbol** is used as a short cut for the address; in most PLC platforms, typing the symbol will automatically call up the address. In the case of a tag-based platform, there may be no address, in which case the symbol is the only address that is used. For tag-based systems, the symbol is actually downloaded into the PLC as part of the program. For address-based systems it usually is not; it is a part of the program saved on the computer, but not present in the CPU. Symbols or tags are usually limited by the number and choice of characters allowed. For example, underscores (_) may have to be substituted for spaces.

A **Description** is purely for the convenience of the programmer. The description is only present in the programming computer and is not downloaded to the PLC. Typically a description can be several lines in length, unlike a symbol. It can usually contain any text character and is used to fully describe aspects of the device or contact, such as its physical location, numerical designation or purpose.

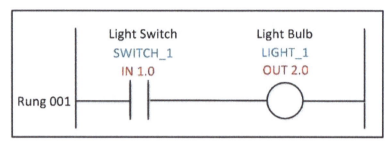

This diagram shows the same circuit as in the electrical diagram on the previous page, but in Ladder Logic. Like the electrical diagram, if the switch contacts on the left are closed, the light will be energized; however the only information you can get from the contact is that it is *Normally Open*.

The device on the left is called a **contact**, and the device on the right is a **coil**. These names come from parts of a relay, which is the basis of Relay Ladder Logic. This type of contact is known as Normally Open, or NO; it is assumed to be not energized or activated in its normal condition. There are also contacts that are in the energized or "on" condition when in its normal condition, these are known as Normally Closed, or NC contacts.

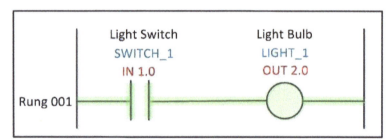

When a contact is viewed in its online condition, it is often highlighted in some way to indicate continuity. In the case of the light switch logic shown before, this indicates that while the switch is Normally Open, in this case something has activated it and the contacts are closed. The green highlight around the coil indicates that it is energized.

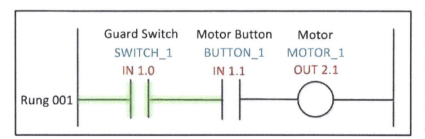

This circuit shows a normally open Guard Switch in series with a normally open motor pushbutton. If the guard switch is a switch on a motor cover, it would be assumed that if the cover is closed, the switch would allow the button to jog the motor. Therefore the switch is actually wired to be closed with the cover on; if the cover is removed, the switch opens, turning off the motor even if the button is being pushed. In other words, if the Guard Switch is closed **AND** the button is pressed, the motor will run. This is a classic example of an AND circuit, two contacts in series. In this case, if the circuit was being monitored through the software, the cover is on the motor, but the button is not being pressed, so the motor is not running.

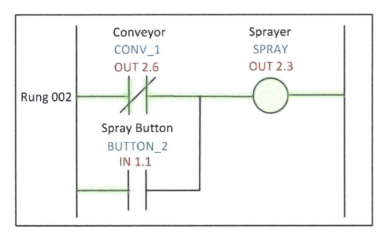

This circuit shows a Normally Closed contact from a conveyor in parallel with a Normally Open button. If the conveyor is on, (that is, if the output for running the conveyor is turned on elsewhere in the program), the sprayer will not be energized, UNLESS the button is pushed. This illustrates a new concept, the **OR** circuit. If the conveyor is off, OR the button is pushed, the sprayer output will activate. In this case, since the coil is activated and the highlighting extends through the conveyor contacts, the conveyor is not running! Note also that it is possible to use output addresses as contacts.

In the previous examples, the coil has indicated the status of the preceding logic. In other words, if there is a continuous path from the left rail to the coil, the coil is on. If not, the coil is off. It is also possible however to **SET** or **RESET** the coils address, making the logic Retentive. This is also sometimes called "Latch" and "Unlatch".

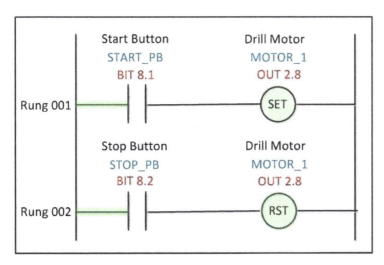

This diagram shows an output that energizes a motor. It can be "latched" on or off by pressing the start or stop button. Notice that the coil is energized, but there is no path or continuity from the contacts to the coils. This means that the output will be maintained until the Stop Button is pressed.

Notice also that in this case a "bit" address is used to control the motor. Since these aren't input register addresses as in the previous examples, that means that they must come from somewhere other than physical inputs. It is hard to tell without doing a cross reference to see where the bit's coil might be, but if a coil for that address can't be found in the program, it likely comes from an HMI, or operator interface.

Contrary to the drawing, it is not advisable to directly latch output addresses. Instead, a memory bit should be latched and unlatched that controls the output.

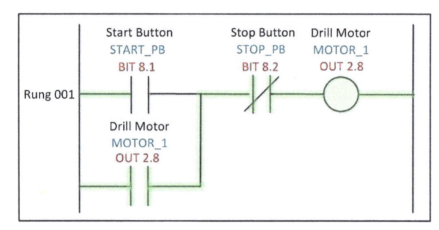

This logic performs exactly the same function as the latching circuit, but in a different way. Sometimes called a "hold-in" or "seal-in" circuit, if the start button is pressed and the stop button is not being pressed, the motor coil will energize. Even if the start button is released, the "hold-in" contact of the motor coil will ensure that the motor stays on until the stop button is pressed. Notice that in this case, the stop button needs to be a normally closed contact, even though both buttons are <u>physically wired</u> normally open in both diagrams.

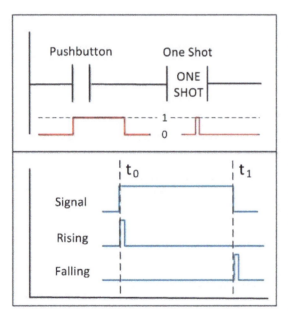

A **One Shot,** single shot or differential pulse is used for a variety of reasons in ladder logic. This diagram shows the result of using a one shot; when a signal changes state the one shot creates a pulse exactly one scan in length. For example, when the Pushbutton is pressed, no matter how long it is held down, the logic will generate one single-scan pulse.

One shot pulses can be generated off of the rising or falling edge of a signal. They are known by different names on different platforms: Allen-Bradley uses OSR and OSF contacts called One Shot Rising and One Shot Falling; ONS one shots are also sometimes used on the rising edge. Siemens calls its one shots Positive and Negative differentials, they look like –(P)- and –(N)- in ladder logic.

Omron's one shots are called DIFU (Differential Up) and DIFD (Differential Down) for the rising and falling edge one shots. They look similar to this: -|DIFU|- and -|DIFD|-

Mitsubishi calls them PLS or Pulse. -|PLS|-

Some types of one shot are placed at the end of the rung as a "box" type instruction. These have two addresses, one for storage and one to be used as the output. These output contacts can be used at multiple places in a program. In-line one shots must each have their own memory bit address; *do not use the same address for multiple one shots!*

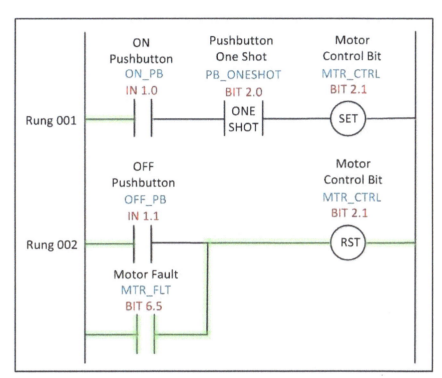

One shots are often used with latching circuits to ensure that logic defaults to the off state as shown. If the one shot is of the "in-line" type described previously, each must have its own address.

As long as the Motor Fault is present, the motor control bit will be held in the off state. To turn the bit back on, the fault will have to be cleared and the ON pushbutton will need to be pressed again.

Notice also that even though the rung is highlighted all the way to the RST coil, the coil itself is not highlighted, indicating that the address is off. Contrast this with the latched coil picture on the previous page; this is typical when monitoring PLC logic with the software.

Exercise 5

1. Draw ladder logic to place a machine into either Auto or Manual mode by pressing physical pushbuttons. Ensure that the machine is placed into Manual if a Fault occurs.

Work Area:

2. Draw logic to accomplish the following:

    a. Turn on a motor using a start button and a stop button using a hold-in contact. Both buttons should be <u>wired</u> Normally Open (NO). Include a NC Fault contact that will stop the motor.
    b. Latch the Fault bit if the motor's overload trips or the guard door is opened while the motor is running.
    c. Turn on a Red Light if there is a fault.
    d. Turn on a Green Light if the motor is running.
    e. Reset the Fault if there is no fault and a Reset Button is pushed.

Assign addresses to all of the contacts and coils using the generic method shown in this document; i.e. BIT, IN, OUT

Work Area:

Timers

The purpose of a timer in ladder logic is to delay the on or off state of a signal. Timers can be used to track the accumulation of time in a process, create a pulse of a fixed length, or determine whether a fault has occurred.

Before looking at how a timer operates, it is important to view its data structure.

Preset	
Accumulated	
Status Bits	...

The Preset and Accumulated values are usually either an integer or double integer value; however Siemens uses a BCD value that incorporates the time base as part of a word (16 bits). The status bits always include a "**Done**" bit, but may also include bits for Timer Enabled or Timer Timing. If the status bits don't include these, they can easily be generated using logic structures.

On Delay:

An On-Delay Timer is used to delay the ON state of its done bit as shown in this diagram.

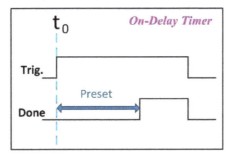

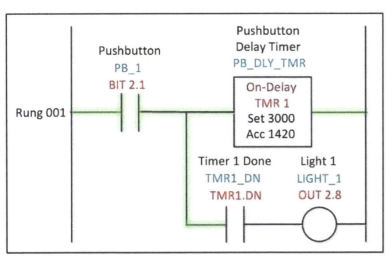

When the button is pushed, the timer begins timing. After the preset time (Set) has expired, the done bit will energize and remain on until the pushbutton has been released. If the Pushbutton is released before the Accumulated time (Acc) reaches the preset, the timer will reset to zero. When the done bit comes on it will energize the output, Light 1.

Tip: Notice that the preset is in thousands. According to IEC 61131-3, a timer counts time in milliseconds, so 3000 is three seconds. Also, the timer will count up as it times. Not all platforms' timers perform this way; some time down, some have a time base that allow them to count in 10ms, 100ms, 1 second or even 10 second increments. Most major manufacturers now have a timer available that meets the IEC definition.

Off Delay:

An Off-Delay Timer delays the OFF state of its done bit. Unlike the On-Delay, the done bit turns on immediately.

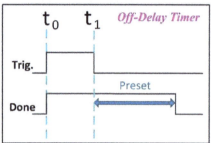

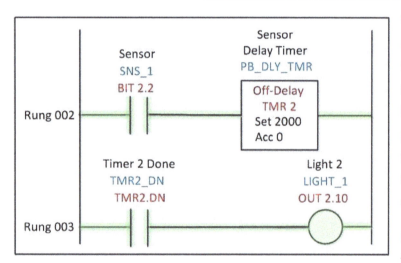

When the sensor is activated, the done bit comes on immediately, therefore the light comes on. When the sensor turns off, the timer begins timing and runs until the preset (Set) value is reached, then the done bit turns off. During this time, the accumulated value is counting up in milliseconds.

Notice that with this type of timer the done contact can't be placed in a branch after the trigger contact as with the on-delay timer. If it was, turning off the sensor would also not allow the light to come on. The off-delay timer can be thought of as a "pulse stretcher".

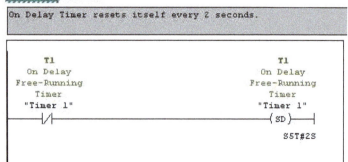

Tip: With many PLCs, the done bit is shown as the same address as the timer. This is a rung or "Network" from a Siemens Step 7 program. Note that the Symbol T1 is on both the SD (On-Delay) coil and the NC contact; the contact is the "done" bit. This rung creates a one scan-length pulse every 2 seconds.

Another aspect of a timer that is not part of its data type is the **RESET** coil. This coil is used to set the accumulated value to zero. All timers have this capability, though the reset is most commonly used in retentive timers.

Retentive On Delay:

A retentive timer keeps its accumulated value even when the energizing contact is not made. This is useful for accumulating run-time on a device or product. This means it must be reset.

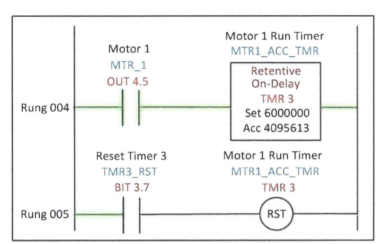

When the motor runs, the timer accumulates time. When it stops, the timer retains its accumulated value and does not reset. When the accumulator reaches the preset value, the timer stops timing, the done bit is energized, signifying that it is time to service the motor.

Usually, rather than placing a high number in the preset, a number such as 60000 (60 seconds) is placed there. The done bit is then used to increment a "Minutes" counter, which resets the timer. When the Minutes counter reaches 60, it increments an "Hours" counter; service on devices is often specified in hours. After the maintenance has been performed, the timer and counters can be reset. *A Siemens Retentive On-Delay Timer acts differently: when the trigger signal is removed, the timer continues timing until done. Siemens timers also time <u>down</u> from their preset.*

Pulse:

Pulse Timers create a pulse of a fixed length when energized; the done bit energizes immediately and stays on until the preset is reached. Some pulse timers' done bits will de-energize if the trigger is removed from the timer early.

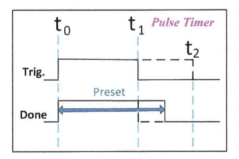

If the trigger signal stays on either shorter (t1) or longer (t2) than the Preset, the pulse output ("done" bit) stays on for the preset time. (Siemens S_PEXT). If the trigger signal is released before the preset time, some pulse timers will turn off the done bit as at t1. (Siemens SE). For PLC brands without a Pulse Timer, the pulse can be created using two On-Delay timers.

Exercise 6

1. Draw ladder logic to delay the start of a motor for 3 seconds. Use a start and stop button to control the motor. Ensure that if the button is released before the motor starts, it will start anyway.

   Work Area:

2. Draw ladder logic to ensure that a sprayer stays on for at least 2 seconds whenever an object passes in front of a photoeye on a conveyor. Ensure that if the conveyor is not running, the sprayer stops.

   Work Area:

3. Draw ladder logic that creates a one second on, two second off train of pulses. Use a pushbutton to sound a buzzer with this signal.

   Work Area:

Counters

A **Counter** is used to add or subtract counts in a register. As with the Timer, a counter also has a data type associated with it.

Preset									
Accumulated									
Status Bits								...	

The Preset is the value at which the done bit will come on in most counters, again with the exception of Siemens. The Accumulated value is the current number of counts in register. Unlike a timer, which stops accumulating when the done bit activates, a counter will continue incrementing. Preset and accumulated values may be an integer or double integer depending on the platform.

Other Status Bits that MAY be present besides the "Done" bit:

a. Count Up (CU): active while up count trigger is active
b. Count Down (CD): active while down count trigger is active
c. Overflow (OV): active when Accumulated value exceeds maximum value for an Integer or Double Integer, depending on brand/platform
d. Underflow (UN): active when Accumulated value exceeds minimum value for an Integer or Double Integer, depending on brand/platform

Tip: The IEC61131-3 definition of a counter states that a counter's done bit will change state when the accumulator reaches the preset. However, Siemens counters, once again, act differently; the done bit is on if the accumulator is over zero. Because of this, Siemens programmers often use a different method to count.

There are three types of counter: Up Counters, Down Counters, and Up-Down Counters. While every brand needs an Up and a Down counter, the Up-Down type is not available on all PLCs.

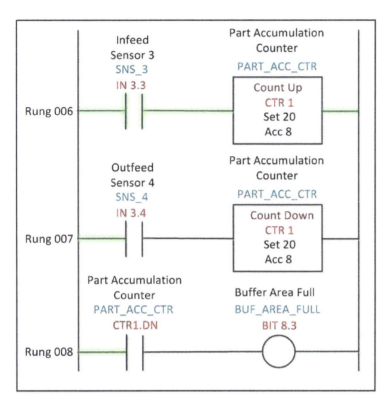

Separate up and down counters can be used with the same effect. As with retentive timers, counters need a Reset bit. Some counters will also use a "Set" bit to place the preset value into the accumulator.

This logic tracks parts into and out of a "Buffer" area on a conveyor system. A sensor is located at the entry and exit of the area, so parts entering make the counter count up, while parts exiting make it count down. When the "Done" bit is active, a gate could be closed preventing new parts from entering the buffer area until parts have exited.

Notice that the same address was used for both counters. This is important to ensure that the same address is counting up and down.

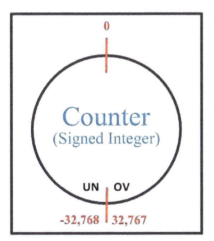

So what happens when a counter reaches the overflow or underflow value? It rolls over to the negative range if the positive value is exceeded, and to the positive side if counting down in the negative direction. This is the purpose of the overflow and underflow status bits, they indicate that this boundary has been crossed. They can be reset (unlatched) after determining what you wish to do about the over or underflow.

If the counter is a Double Integer based device, the limits are −2,147,483,648 to +2,147,483,647.

The reset for the counter works the same as the one shown in the previous timer diagrams:

Exercise 7

1. Draw ladder logic that counts parts into a box. When the box is full (10 parts), latch a memory bit that controls a light, the light tells an operator to remove the box. Use a sensor to detect that the box has been removed. When the box has been removed, reset the counter and the memory bit that controls the light.

Work Area:

2. Draw ladder logic that uses a Retentive On-Delay timer to accumulate run time on a motor. When the timer reaches one minute, use the done bit to increment a "Minutes" counter and reset the timer. When the counter reaches 60 minutes, use its done bit to increment an "Hours" counter and reset the Minute counter. When the Hour counter reaches 10,000, it is time to perform maintenance on the motor.

   Do not reset the Hours counter automatically; use a pushbutton that is pressed by the maintenance technician, confirming that maintenance has been done. How can you incorporate logic to ensure that the technician really did do the maintenance?

   Work Area:

   Don't forget to reset the timer and other counters when the operator presses the button!

Data and File Movement

An important part of programming is the manipulation, modification and movement of data. The simplest method of doing this is to simply move a number from one location or register to another.

Move:

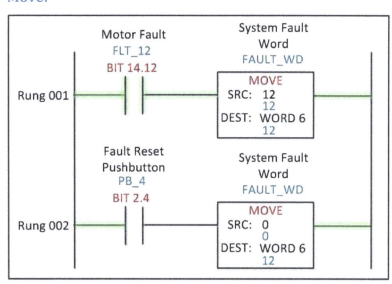

Despite the common term "Move" that is used for this operation, it is actually a copy, since the value also remains where it was moved from.

This logic shows an integer being moved into a register based on a numbered fault. The number can be used to display a message on a touchscreen or in a comparison to set the fault status by latching a bit. The numbers in blue show the actual number in the register (Word 6) if the logic was being monitored online.

In this example, constants are being moved, but numbers can also be moved from one register to another also.

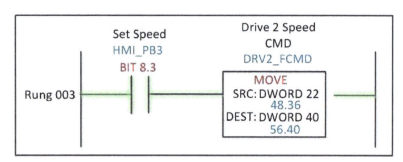

This logic shows a speed being moved into a VFD speed command. The numbers are floating point or REALs, so they require a double word, or 32 bits for the register size.

Masked Move and Shift:

Parts of numbers can also be moved. If it is necessary to extract a specific part of a 16 or 32 bit number, a **Mask** can be used to move only that part, as shown in the diagram.

The Source is a 32 bit number, which is four bytes. If one wanted to move only the second byte into a register, the mask would contain ones wherever the data to be moved, and zeroes elsewhere. Masks are usually entered as a Hexadecimal number, which in this case would be 00FF0000, or just FF0000.

Of course, then you have the problem that the byte in the destination is in the wrong location within the double integer, this is easily remedied by using a **Shift** instruction. The result is shown below. In this case the shift is to the right sixteen spaces.

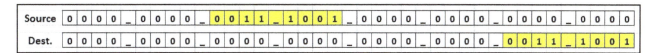

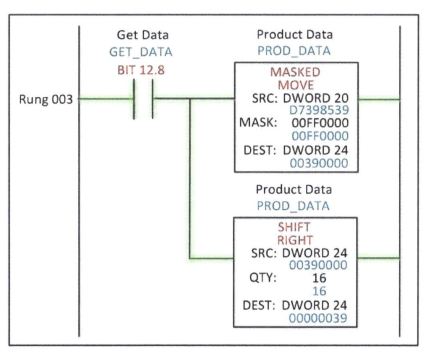

The ladder logic for a Masked Move and a Shift are shown here. The shift instruction may not have a destination as in this diagram, however it is shown here to indicate the number before and after the shift. Shift Left instructions can also be used.

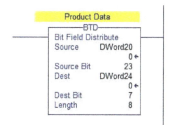

Tip: The Allen-Bradley ControlLogix 5000 platform has an instruction that does both of these functions at once (Move and Shift), called "Bit Field Distribute" or BTD.

File Copy:

File instructions are used to move data structures that are larger than a single element or 32 bits. If a file is a single structure, such as a UDT or an individual data type such as an Array, a simple "Copy" type command can be used. If it is necessary to move specific overlapping sections, more complex pointer-based commands may be used.

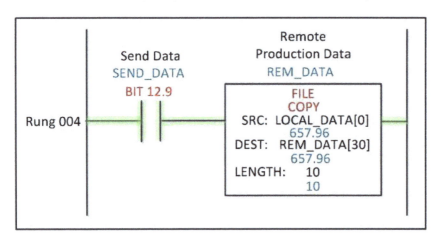

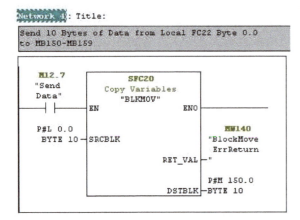

Tip: This logic illustrates the use of pointers in Siemens S7 platform. SFC20 is a Block Move command that allows the use of pointers to designate the type of data (Local and Marker Memory Bytes), the size of the data (10 Bytes) and the location within the structure (0.0 to 150.0). This command allows movements to be specified to the bit level.

The RET_VAL output is present on many Siemens instructions to allow for error return values.

Comparisons

Data comparison is an important element of PLC programming. Though comparisons deal with numerical values, they are input type instructions. That is, they evaluate to either true or false.

Standard comparison instructions that are found in any PLC include Equal (=), Not Equal(<>), Greater Than (>), Less Than (<), Greater Than or Equal (>=), and Less Than or Equal (<=).

Some PLC platforms require that the data type being compared is the same, while others allow comparisons between different types. Where data types must be the same, a conversion instruction may be required.

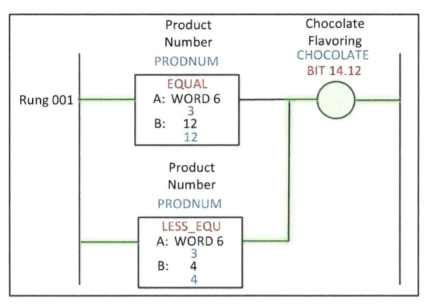

This illustrates the use of two comparison instructions used to activate a bit that is used to add chocolate flavoring to a recipe. If the selected product number is between 0-4 (inclusive) or equal to 12, the bit will be true. In this case the selected product number, in Word 6, is 3.

Comparisons are also often used in automatic sequences to move from one step to another, or to determine when an output will be activated.

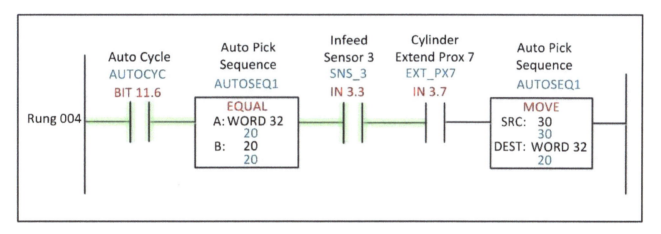

This logic increments a sequence to the next step if the machine is in Auto Cycle mode and a couple of sensors are activated. This only happens if the sequence is in step 20.

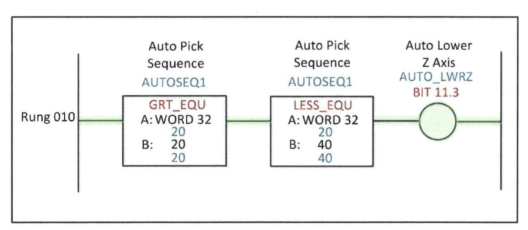

Comparison instructions can be placed in series to form a "window" during which a statement is true. In this case, the command to lower the Z Axis of a pick-and-place mechanism will be on if the sequence is in steps 20-40, inclusive.

Some PLC platforms have instructions that form this window in one instruction.

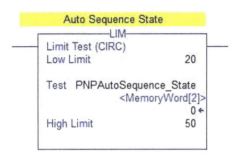

Tip: The Allen-Bradley "Limit" instruction can be used to form a window where a sequence state is tested between a low and high value. If the Low Limit value is greater than the High Limit value, the instruction operates in reverse, where if the tested value is outside of the range, the instruction is true!

A Mask can also be used with an equal instruction on some platforms. As with the Masked Move instruction, wherever there are ones in the mask, the values will be compared.

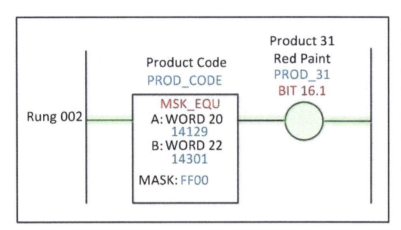

In this logic, it appears that the two values in Word 20 and Word 22 are different, yet the coil is energized. This is because only the 8 most significant bits are being looked at, while the lowest 8 bits are ignored.

The numbers are equal!

This is what the numbers 14,129 and 14,301 look like in Binary.

A:	0 0 1 1	0 1 1 1 1	0 0 1 1	0 0 0 1
B:	0 0 1 1	0 1 1 1 1	1 1 0 1	1 1 0 1
Mask:	1 1 1 1	1 1 1 1 1	0 0 0 0	0 0 0 0

 Exercise 8

1. If the Mask in a Masked Equal Instruction is 00FF, are 31,290 and 4410 Equal? _____

Work Area:

2. Write Ladder Logic that increments an Auto Sequence through three different steps based on I/O and/or internal bits. On the last step, reset the sequence to zero.

Work Area:

Why do you think the Auto Sequence examples in this chapter increment by 10, rather than 1?

_____

Math Instructions

Processing of data in a PLC often involves performing mathematical operations on data. Not all processors allow math to be performed between different data types, so it may be necessary to convert one data type to another.

Conversion

Common conversions include the following:

1. **Integer to Double Integer, Double Integer to Integer**. The inherent problem in this type of conversion is that the number in the DINT won't fit into the INT. Remember that these are usually signed values, so the largest value you can put into an integer is -32,768 to 32,767.

2. **Double Integer to Real, Integer to Real**. Some platforms will have DINT to REAL but not INT to REAL. In this case the integer will have to be converted to a double integer first.

3. **Real to Double Integer, Real to Integer**. In this case you will lose the value after the decimal point. Again, not all platforms have Real to Integer conversion.

4. **Integer to BCD, BCD to Integer**. Remember that your BCD value will contain more bits than your integer.

Platforms in which data conversions are necessary include Siemens and Koyo. BCD data conversions are possible on most platforms.

Addition and Subtraction

Addition and subtraction are commonly used for all data types. A typical use might be to increment or decrement a register by some amount:

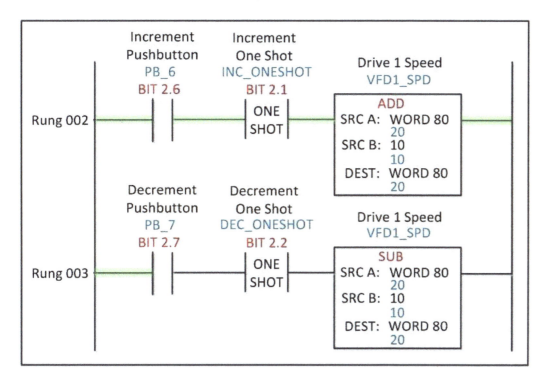

In this logic, pushbuttons are used to either add or subtract 10 from a register each time the button is pushed. Notice that the number is subtracted from a register and then placed back into the same register; this is the normal way to accomplish this. Also notice that a one-shot is used on the pushbuttons. If the one shot is not used, the instruction will not add or subtract every time the button is pushed, instead it will add or subtract *every scan* while the button is being pressed! You could end up with a very large number in Word 80 if you aren't careful!

A common name for Rung 002 is an **Accumulator**, while the following rung is sometimes known as a **Decumulator**.

Tip: A Siemens counter has quite a few limitations. The "Done" bit is on if the counter is not at zero, and the preset and accumulated values are signed BCD numbers; they only count from -999 to +999.

Because of this, Siemens programmers will often use Accumulation/Decumulation logic to count. In this case the "Done" bit will be created by using a Greater Than or Equal instruction, the Reset command moves zero into the accumulator value, and the registers would increment and decrement by 1.

Multiplication and Division

As with addition and subtraction, multiplication and division can be done with any data type, however greater care must be taken for several reasons.

If you multiply two integers, it is important to ensure that the result will fit into the destination address. For instance, if 20,000 is multiplied by 20,000, the answer of 400,000,000 will not fit into an integer register or tag.

When dividing, if the denominator is too small (or zero!) the same thing will happen.

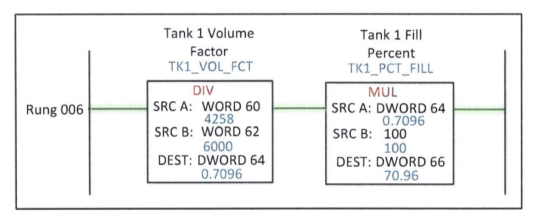

This logic calculates the percentage of fill of a 6,000-gallon tank. The measured volume (WORD 60) is divided by the total volume (Word 62), and then multiplied by 100. In this example, data types are mixed; an integer is divided by an integer and the result is placed in a REAL. The REAL is then multiplied by an integer and the result placed in another REAL. If the PLC being used did not support this functionality, data conversions from INT to REAL would be necessary.

On some brands of PLC, the data types must be the same to perform math. In this case the numbers must be converted as shown in the Siemens S7 example below.

Network 6 : Convert ACC

Convert Timer 4 BCD Value from BCD to REAL

```
         BCD_I                              I_DI                          DI_R
      EN      ENO                        EN      ENO                   EN      ENO

  MW40              MW42       MW42              MD44       MD44              MD48
 Timer 4          Timer 4    Timer 4          Timer 4    Timer 4          Timer 4
 Acc Value        Acc Value  Acc Value        Acc Value  Acc Value        Acc Value
 "TMR4_ACC" —IN     INT        INT               DINT       DINT             REAL
                  "TMR4_ACC_  "TMR4_ACC_        "TMR4_ACC_ "TMR4_ACC_       "TMR4_ACC_
            OUT —I"      I" —IN    OUT —DI"           DI" —IN    OUT —RL"
```

Network 7 : Calculate Percent

Calculate Accumulated percent of Preset

```
         DIV_R                              MUL_R
      EN      ENO                        EN      ENO

  MD48              MD56       MD56              MD60
 Timer 4          Timer 4    Timer 4          Timer 4
 Acc Value        Base       Base             ACC
 REAL             Percent    Percent          Percent
 "TMR4_ACC_       Value      Value            Time
       RL" —IN1   "TMR4_     "TMR4_           "TMR4_ACC_
            OUT —BASE_PCT"  BASE_PCT" —IN1 OUT —PRE_PCT"
  MD52
 Timer 4                    1.000000e+
 Preset                          002 —IN2
 Value REAL
 "TMR4_PRE_
       RL" —IN2
```

As in the previous example, this logic calculates a percentage, this time a timer's percent of completion. Siemens' timer accumulated values are in BCD, so the number is converted from BCD to Integer to Double Integer to REAL. Siemens' timers also time from the setpoint down to zero, so the final value would need to be subtracted from 100% to present the remaining time to completion (100.0 – TMR4_ACC_PRE_PCT, or 100 - MD60).

Exercise 9

1. A Variable Frequency Drive (VFD) is used to control a conveyor. Its maximum speed is 1750 RPM, but the number sent from the drive is in integer form. At full speed, the integer reads 31,760, while it reads zero when stopped. Write ladder logic to calculate the percentage of a drive's actual speed to its total speed, also providing a REAL speed in RPM.

Work Area:

2. Production and reject data from a manufacturing line is entered manually at the end of each shift. After completion of the third shift, there are three registers named Shift1_Prd, Shift2_Prd and Shift3_Prd containing the total parts made that shift, and 3 more registers named Shift1_Rej, Shift2_Rej and Shift3_Rej that contain the number of failed parts. Write ladder logic to calculate Total Parts, Total Rejects and Total Good Parts for the day.

Work Area:

Scaling

An important mathematical function is that of converting raw analog values into usable units of measure, or converting one unit into another. This is known as **Scaling**, and it follows a standard formula, y =mx+b. **Y** is the units of the Y axis, and **X** is the units of the X axis. **B** is known as the **Offset**, while **M** is the **Scalar**, determined by dividing the "rise", or increase of the Y axis, by the "run", or increase of the X axis.

As an example, let's look at the graph below:

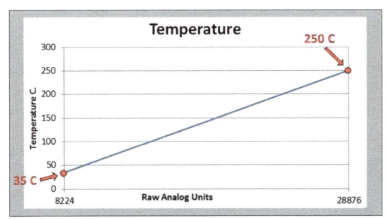

A temperature sensor produces a 0-10v signal. This is wired into an analog card which produces a signal that ranges from 0-32,767, a signed integer.

A thermometer is used to measure the actual temperature at two different points and the raw value is recorded for each measurement; the first point P1 is recorded as 8,224 on the analog card at 35 degrees C, while the second (P2) is recorded as 28,876 at 250 degrees C.

The first step in scaling the raw measurements into degrees is to calculate M, the Scalar. The "rise" or difference in Y values is Y2-Y1, 250-35, or 215. The "run" or difference in X values is X2-X1, 28,876-8224, or 20,652. Dividing the rise by the run produces a Scalar M of 0.01041061.

The next step is to calculate the offset B. Since y=mx+b, the B factor can be calculated as B=Y-MX. Substituting the values for P1, which were Y1 and X1, the calculation becomes B = (35-(0.01041061*8,224)), which yields an offset of -50.6168894. These two constants, M and B can be used to calculate Y for any inserted value of X.

As an example of how to use this formula in calculating a temperature, assume that the temperature sensor reads a value of 10,512 into the analog card. If the formula y=mx+b is used, the temperature is (0.01041061*16,512) – 50.6168894, or 121.28 degrees C.

This formula can also be used to calculate all of these variables in one group of calculations. Allen-Bradley's Scale with Parameters (SCP) instruction allows the programmer to enter Raw High and Low and Engineering High and Low units into the instruction along with the measured value from the card. It then outputs the scaled value to another variable.

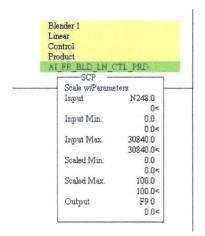

The input value is N248:0, a signed integer value from an analog card. The Minimum and Maximum Input values are taken from observed values coming from the card, while the Minimum and Maximum Scaled values of 0.0 and 100.0 represent 0 to 100 percent of the Input values. The result is moved into F9:0, a Floating Point or REAL register.

Unfortunately, many PLCs do not have this instruction, including A-B's ControlLogix, but the math can easily be reproduced as shown in the following logic:

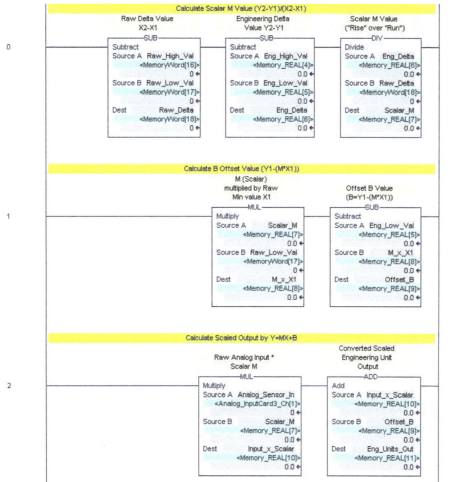

This logic can also be packaged inside of a subroutine, function or "Add On Instruction" (AOI) where parameters can be passed into it and the results passed out.

The first two rungs use High and Low variables passed "by reference", (as shown in the SCP instruction), to calculate the internal (local) variables M and B. These are then used in the third rung to scale the Analog Sensor Input to Engineering Units Out.

Exercise 10

1. A tank used for blending juice holds approximately 8,000 gallons of liquid. There is a pressure transducer that produces a 4-20mA signal, it is wired into channel 1 of an analog card.

   The tank is filled with 6,000 gallons of juice; the reading of the analog card is recorded as 24,780. The tank is then drained completely and the value from the transducer is recorded as 96.

   Draw ladder logic that scales the raw reading from the transducer into gallons. There are 3.78541 liters in one gallon. Also calculate the number of liters.

Work Area:

Advanced Math

In addition to the addition, subtraction, multiplication and division instructions mentioned previously, here is a list of more advanced math instructions along with their purpose:

Exponent – An exponent signifies the number of times that a number is to be multiplied by itself. The exponent is usually indicated by using the "^" sign, so 3^4 is 3 x 3 x 3 x 3.

Logarithm, Natural Log (LOG, LN) – A logarithm (LOG) is the inverse of an exponent. For instance, if 2^3 is 2 x 2 x 2 = 8 where 3 is the exponent, then the LOG of 8 in Base 2 is 3. Where a LOG is typically base 2, a Natural Log (LN) has a base of 2.718. Natural Logs are often used in math and physics (calculating decibels and pH), while LOG is often used for computer calculations.

Sine, Cosine, Tangent (SIN, COS, TAN) – Also known as Trigonometric functions, these are used to calculate geometrical coordinates. These functions -- along with their reciprocals Cosecant, Secant and Cotangent and their inverses Arcsine, Arccosine and Arctangent -- are often used in motion control applications.

Modulo (MOD) – This function calculates the remainder after a division operation.

Absolute Value (ABS) – Returns the positive version of a number even if it is negative. The Absolute Value of both -15 and 15 is 15.

Other Instructions

There are a wide variety of other instructions available on different PLC platforms in addition to those listed previously. These are just a few that are common to some of the major brands of PLC.

String Operations

As mentioned in the data section of this manual, strings are arrays of SINTs, or Single Integers (Bytes). The array elements contain ASCII characters, which can be thought of as printable characters with a few non-printable commands included. Values contained in strings can be displayed as decimal or hexadecimal numbers, or as text characters. If in text, they are often displayed with a "$" sign before the character, such as Text = $T, $e, $x, $t characters. These equate to the decimal numbers 84, 101, 120, 116 or the hex numbers 54, 65, 78, 74. These can be found in a standard ASCII table; there is one in the appendix of this manual.

Strings may also contain a length (LEN) field that contains the number of characters that exist in the string. For instance, if a string has space for 80 characters, but is filled with the characters "Today is Tuesday, September 13" then LEN = 30.

Concatenate (CONCAT) – Connect two strings together, one after another.

Middle (MID) – Copies a specified String into the middle of another String at a specified location.

Find (FND) – Locate the starting position of a specified String within another String. Usually returns the position of the found String.

Delete (SDEL) – Removes characters from a String at a specified position.

Insert (INS) – Adds characters to a String at a specified position.

Length (LEN) – Finds the number of characters in a String if length is not part of the string definition.

PID Instructions

PID, or Proportional-Integral-Derivative instructions control a process variable such as flow, pressure, temperature or level.

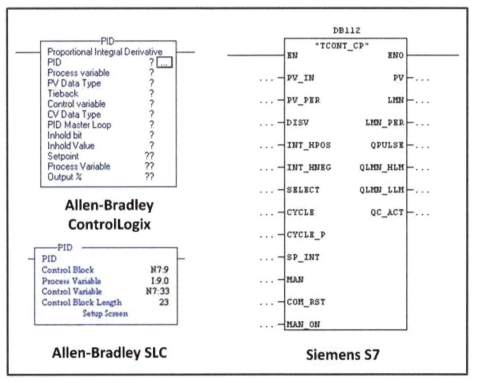

Allen-Bradley ControlLogix

Allen-Bradley SLC

Siemens S7

There are a wide variety of parameters that can be set for control. These may be passed in as variables as shown in the ControlLogix and S7 diagrams, or set up on a special screen as shown for the SLC.

Motion Control Instructions

Newer PLC platforms may use multi-axis controller cards to coordinate movement. There are many possible commands associated with controlling axes: start, stop, jog, and direction are just a few. Numerical values such as speed, acceleration and deceleration and "go to position" commands are also common.

In addition to these individual axis commands, many coordinated movement commands are also included.

Allen-Bradley's RSLogix5000 software (ControlLogix and CompactLogix) includes 6 folders with 43 different instructions relating to motion control. These are available in Ladder and sometimes Structured Text or Function Block:

Motion Configuration – 4 Instructions. Related to tuning and parameter assignment for axes.

Motion Event – 6 Instructions. Related to arming and disarming events in the motion control card.

Motion Group – 4 Instructions. Related to issuing simultaneous command to multiple axes.

Motion Move – 12 Instructions. Related to individual issuing of motion commands to an axis.

Motion State – 8 Instructions. Related to directly controlling the operating state of an individual axis.

Multi-Axis Coordinated Motion – 9 Instructions. Related to controlling multiple axes in coordinate movement as with robotics. XYZ, XY, Articulated and SCARA configurations are all addressed.

Communications Instructions

Communications instructions are used to access a port in order to send or receive data.

Siemens Step7 software contains a range of System Functions and System Function Blocks that are used to send and receive data in a variety of ways. These are chosen based on the type of data, type of communications and, in some cases, the protocol used.

Many of these system blocks are available in the CPU; they just need to be called as a routine. Different parameters need to be filled in on the block. Most of these system blocks are categorized as COM_FUNC (Comm. Function), DP (Profibus), PROFIne2, (Profinet) and occasionally TEC_FUNC for ptp, a Siemens point to point or RS422 protocol.

Here is a list of system blocks and abbreviations for Siemens, please look at your help files for details:

BLOCK#	NAME	TYPE		BLOCK#	NAME	TYPE
SFB8	USEND	COM_FUNC		SFB104	IP_CONF	COM_FUNC
SFB9	URCV	COM_FUNC		SFC7	DP_PRAL	DP
SFB12	BSEND	COM_FUNC		SFC9	EN_MSG	COM_FUNC
SFB13	BRCV	COM_FUNC		SFC10	DIS_MSG	COM_FUNC
SFB14	GET	COM_FUNC		SFC11	DPSYC_FR	DP
SFB15	PUT	COM_FUNC		SFC12	D_ACT_DP	DP
SFB16	PRINT	COM_FUNC		SFC14	DPRD_DAT	DP
SFB19	START	COM_FUNC		SFC15	DPWR_DAT	DP
SFB20	STOP	COM_FUNC		SFC60	GD_SND	COM_FUNC
SFB21	RESUME	COM_FUNC		SFC61	GD_RCV	COM_FUNC
SFB22	STATUS	COM_FUNC		SFC62	CONTROL	COM_FUNC
SFB23	USTATUS	COM_FUNC		SFC65	X_SEND	COM_FUNC
SFB31	NOTIFY8P	COM_FUNC		SFC66	X_RCV	COM_FUNC
SFB33	ALARM	COM_FUNC		SFC67	X_GET	COM_FUNC
SFB34	ALARM8	COM_FUNC		SFC68	X_PUT	COM_FUNC
SFB35	ALARM8P	COM_FUNC		SFC69	X_ABORT	COM_FUNC
SFB36	NOTIFY	COM_FUNC		SFC72	I_GET	COM_FUNC
SFB37	AR_SEND	COM_FUNC		SFC73	I_PUT	COM_FUNC
SFB52	RDREC	DP		SFC74	I_ABORT	COM_FUNC
SFB53	WRREC	DP		SFC87	C_DIAG	COM_FUNC
SFB54	RALRM	DP		SFC99	WWW	COM_FUNC
SFB60	SEND_PTP	TEC_FUNC		SFC103	DP_TOPOL	DP
SFB61	RCV_PTP	TEC_FUNC		SFC112	PN_IN	PROFIne2
SFB62	RES_RCVB	TEC_FUNC		SFC113	PN_OUT	PROFIne2
SFB73	RCVREC	DP		SFC114	PN_DP	PROFIne2
SFB74	PRVREC	DP				
SFB75	SALRM	DP				

Allen-Bradley's messaging is generally handled by the MSG instruction:

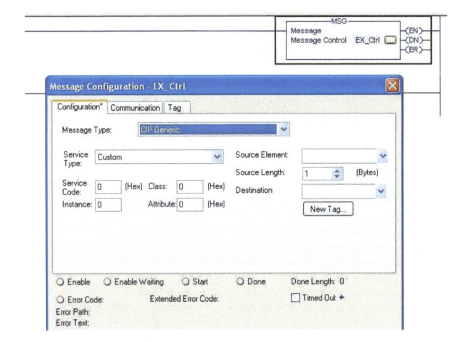

This instruction requires a message control tag specified at the controller level, shown as EX_Ctrl in this image.

After defining the control tag, a message configuration screen is accessed and the type of communications and, ideally, the path to the remote device are specified.

Protocols for Ethernet, DH485, DH+ and Serial DF1 along with SERCOS to motion control devices are available. The target node can be specified as PLC2, PLC3, PLC5, SLC and Generic CIP devices.

There are also several ASCII serial port instructions available for reading, writing, handshaking and buffer control on a basic level.

Program Control Instructions

Program control includes jumping to or calling subroutines, disabling parts of a program by jumps or "MCR" commands, looping by jumping or by using "For/Next", or redirecting the program flow in other ways. Following are some of the more common program control instructions:

Jump Subroutine/CALL – These instructions redirect the program scan to the start of a subroutine or function. At the end of the called subroutine, the flow is redirected to just after the jump or call statement.

Jump/Label – These instructions redirect the scan to a labeled point in the same routine. If jumping forward, some code will not be executed. If jumping backward, code within the zone will be executed over and over (looping) until redirected by another jump, usually associated with a counter which is preset with the number of loops to be executed.

End/Temporary End - This ends the scan of the routine and does not scan code past that point. This is often conditional, controlled by a BOOL or other logic.

For/Next, Do While – Similar to a loop as described above, a For/Next instruction is usually set to operate a specific number of times. A Do/While statement executes at least once, and remains active until a defined condition is met. Both of these instructions are seen most in Structured Text. If used in Ladder, care must be taken not to exceed the Watchdog Timer.

Master Control Relay (MCR) – This instruction is used in pairs. If the first instruction is true, the program proceeds normally. If false (not activated), the physical outputs within the MCR zone will be de-activated. This is not true for outputs that are latched on. ***The MCR instruction should not be used to replace hardware MCRs.***

Miscellaneous/Other Instructions

There are a wide variety of other instructions available, far too many to list here. Every manufacturer has its own instruction set and different names for the instructions.

Here a few general categories of instructions with their uses:

LIFO and FIFO Instructions – LIFO is an acronym for "Last In, First Out", while FIFO stands for "First In/First Out". These instructions operate on a "Stack", which can be configured two different ways.

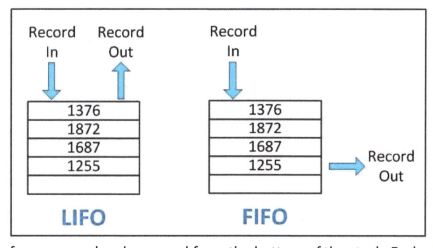

The first stack is similar to a plate dispenser at a restaurant; imagine a spring at the bottom of the stack that pushes items up as they are removed. Values are entered and removed from the top.

The second stack or FIFO allows values to be entered from one end and removed from the bottom of the stack. Each of these stacks have multiple instructions that may be used to manipulate the values. The major instructions are **Load**, which places a new value or record on the stack, and **Unload**, which removes a record or value.

Sequencer Instructions – A sequencer, sometimes called a "drum sequencer", monitors and controls repeatable operations. These instructions also use a stack, but the numbers in the stack are treated as binary values that represent conditions or drive outputs.

A Sequencer Input (SQI) instruction is used to detect when conditions are correct to index the sequencer. If the bit pattern in the designated register matches that of the next position in the sequencer, the sequencer's position value register will increment by one. The bit pattern often represents physical input states.

The Sequencer Output (SQO) instruction is used to set output conditions. These are also represented by a bit pattern, often mapped to physical outputs. The SQI and SQO instructions are usually used in pairs, with the SQI dictating the conditions that index the SQO.

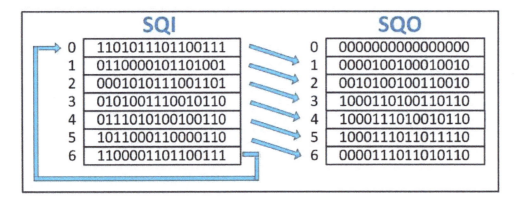

The Sequencer Load (SQL) instruction can be used to place values (bit patterns) into the sequencer's register. This is much like a "teach" function; the instruction looks at a register, often representing inputs. When the SQL is executed, the pattern is loaded into the next location in the stack.

Sequencer Compare (SQC) is another instruction that is sometimes used in order to index the position number of a sequence.

Statistical Instructions – There are various instructions available on some platforms to perform statistical math, such as standard deviation, moving averages and finding minimum and maximum signals in a specified period of time.

Other Instructions are available for Safety Functions, Signal Filtering, VFD (Drive) Control, Equipment Phasing (State Programming) and many others. For full knowledge of a specific platform, it is a good idea to read the programming manual and help files for instructions. Don't forget that many of these require the use of languages other than ladder! It is also possible to build these instructions yourself by using Add-On Instructions or Functions.

Exercise 11

1. What kind of applications are Trigonometric functions used for? _____

    _____

2. Decode the following hexadecimal ASCII characters using the table in the appendix:

    47 ___ 6F ___ 6F ___ 64 ___ 20 ___ 4A ___ 6F ___ 62 ___ 21 ___

    _____

3. Can a JUMP instruction be used to move backwards in a program? _____

4. What do the acronyms "FIFO" and "LIFO" stand for? _____

    _____

5. What is the purpose of a Sequencer instruction? _____

    _____

    _____

Maintenance and Troubleshooting

There are a number of tools and techniques common to all PLC platforms that can aid a technician in isolating the causes of problems. An important thing to remember when using these tools is that *The PLC's program cannot change without someone changing it!* Programs can't change themselves, they either run or they don't.

Forcing

One method of determining whether an input or output is working properly is using a **Force**. In the case of an input, it is obviously not possible to physically force a point on a card. You would have to place a voltage on the point in order to energize it. So what are you forcing when you force an input? *Only the input table*. This means that when you apply the force, the contacts or values related to that point will change only in the program, not on the card itself.

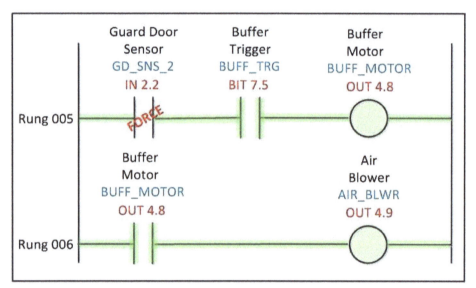

In this case, the Guard Door Sensor is not allowing the Buffer Motor to run even though the trigger signal is shown to be active. A force is applied to input 2.2 and the motor runs. This is a verification that the electrical signal to the input needs to be checked; maybe it is a bad sensor, maybe the wire is disconnected, or maybe the input point on the card itself is bad. In this case, the forcing of the input has helped to determine the problem. After the problem has been isolated, the force can be removed and the problem fixed. *The force should not be used to hide the problem!*

On most PLC platforms, there will be some kind of indication on the contact showing that the input or output is forced. There is also usually a light on the processor itself indicating that a force is present.

Forcing an output is the _exact opposite_ of forcing an input. If a force is placed on the output point, the physical output will be energized, but _the output image table will not be affected_.

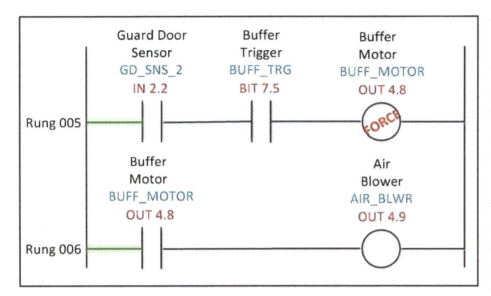

The logic energizing the Buffer Motor is not true, but a force is applied to the output. The physical output comes on, and the Buffer motor runs. Note, however, that the Air Blower that usually comes on whenever the motor runs is not energized. This is because the image table is not affected by the force. The forced coil is actually transparent to the logic; if conditions are true up to the coil, the image table will be updated and the Air Blower coil will be energized.

In most PLCs only inputs and outputs can be forced, but some platforms allow the forcing of memory also.

Installing and activating forces is generally a multi-step process. The force is installed in one step, and then activated as a separate action. This is because **forces can be very dangerous if implemented incorrectly**; you are telling the PLC to do something unnatural, outside of its coding. A force can, however, be helpful in the troubleshooting of a system.

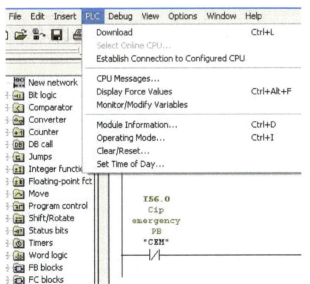

The image to the left shows how forces are accessed in Siemens Step 7 software. From the editor, a Force table is opened from the PLC menu. Force addresses are entered into the table, and then activated.

When active, a red "F" appears by the address indicating that a force is being used.

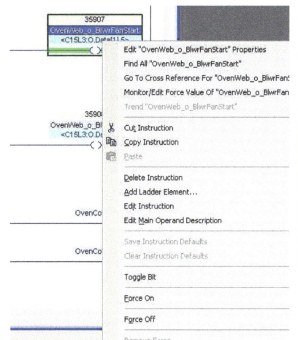

In Allen-Bradley's software, forces are installed by right clicking on an address in the program. After installing the force, forces must be enabled using the dialog shown below:

This two-step process ensures that the programmer truly intends to create a force.

Searching and Cross-Referencing

In order to diagnose problems in a machine, it may be necessary to trace logic through a program. There are a number of tools available to determine the location of addresses, check whether addresses have been used, and substitute one address for another. Searches can also locate words in the comments of a program.

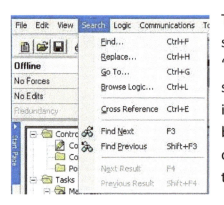

There is usually a tab in the software that will allow various "search" or "Go To" options as shown. Right clicking an address in the program will also often bring up a selection allowing other instances of the address to be found.

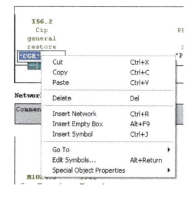

One of the most useful tools when trying to determine why an output is not being activated is the **Cross-Reference**. A cross-reference shows all of the places in a program where an address has been used.

Usually, troubleshooting starts with finding the coil of an address. Right clicking the address brings up a selection for cross-reference or "Find All"; this in turn brings up a list of all of the places where the address is used in the program.

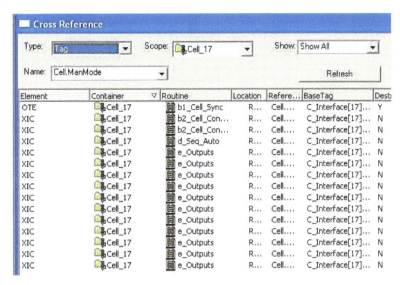

Selecting the location of the coil takes you to the rung or network where the coil (OTE) is activated. With proper programming technique, there should only be one place where a coil will be located for any address!

Addresses can then be traced from rung to rung until finally the cause of the problem can be identified.

The following rungs illustrate tracing the cause of an output failing to come on:

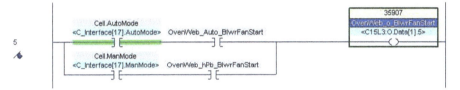

This rung shows that the Oven Web Blower output (the coil) is not energized. Since the

AutoMode contact is energized, a cross-reference of OvenWeb_Auto_BlwrFanStart, is executed. Searching for the coil brings up this rung:

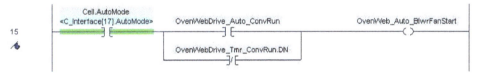

This in turn prompts the searcher to look for

OvenWebDrive_Auto_ConvRun:

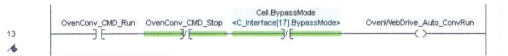

Which in turn brings up

OvenCnv_CMD_Run:

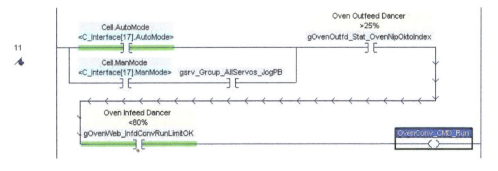

And on to the next location on the following page:

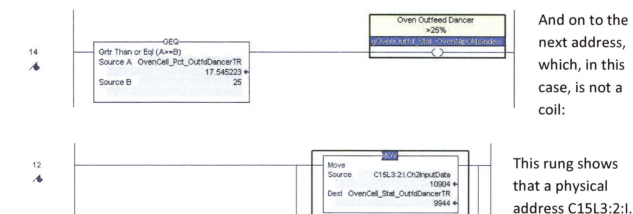

And on to the next address, which, in this case, is not a coil:

This rung shows that a physical address C15L3:2:I. Ch2InputData (an analog input value) is moved into the variable that we are looking for. Cross-referencing and searching will always end at either a physical input point, or at an address with no coil or address that has been changed by the program. This last would mean that the signal comes from outside the controller, such as an HMI, SCADA or even a signal from another PLC.

Did you notice the little flags next to the rungs? Allen-Bradley's ControlLogix platform has a toolbar called "Bookmarks" that allow a programmer to mark rungs and then index through them. Very handy!

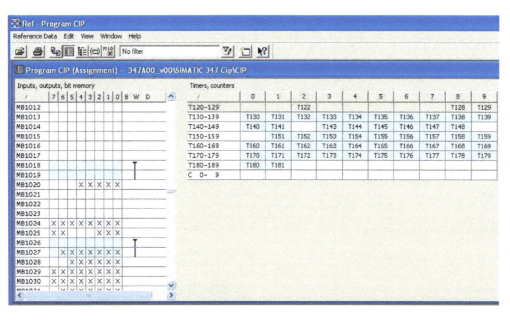

Siemens also has a useful tool that allows one to look at which registers have been used. This can also be used to find addresses with no symbol or symbols with no address. Notice that several bits have been assigned that are also a Word address (MB1026 & MB1027). This can be useful to spot interferences.

Exercise 12

1. Forcing an input applies voltage to the physical input. True: ___ False: ___
2. Forcing an output applies voltage to the physical output. True: ___ False: ___

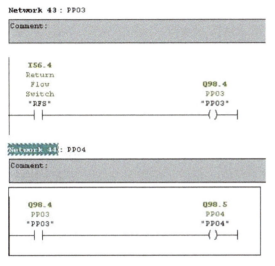

3. In the diagram at left, if Q98.4 "PP03" is forced, what will the state of Q98.5 "PP04" be? (Assuming that input I56.4 is off)

4. If the force on Q98.4 is removed and I56.4 is forced, what will the state of Q98.5 be?

5. To trace a signal through a program, what types of instructions should you look for?
_____

6. What two types of addresses will represent the end of a search or cross reference?
_____
_____

Major Platforms

Following is a list of major PLC manufacturers, website and their country of origin.

ABB – Switzerland   http://new.abb.com/plc

Allen-Bradley – USA   http://ab.rockwellautomation.com/Programmable-Controllers

Automation Direct-USA   https://www.automationdirect.com/

B&R – Germany   http://www.br-automation.com/en-us/products/control-systems/

Beckhoff – Germany   http://www.beckhoff.com/

Bosch – Germany   http://www.boschrexroth.com/

GE Fanuc – USA   http://www.geautomation.com/products/programmable-automation-controllers

Hitachi – Japan   http://www.hitachi-ies.co.jp/english/products/plc/index.htm

Idec – Japan   http://us.idec.com/Home.aspx

Keyence – Japan   http://www.keyence.com/products/controls/plc-package/index.jsp

Koyo – Japan   http://www.koyoele.co.jp/english/product/plc/

Mitsubishi – Japan   http://www.mitsubishielectric.com/fa/products/cnt/plc/

Omron – Japan   https://industrial.omron.us/en/products/programmable-logic-controllers

Panasonic – Japan   https://na.industrial.panasonic.com/products/industrial-automation/factory-automation-devices/programmable-controllers

Siemens – Germany   http://w3.siemens.com/mcms/programmable-logic-controller/en/Pages/Default.aspx

Toshiba – Japan   http://www.toshiba.com/tic/industrial-systems/plcs

Others (smaller brands or may be obsolete): Cutler-Hammer, Eagle-Signal (Eptak), Giddings & Lewis, Philips, Square D, Texas Instruments, Triconex, Velocio, Vipa, Westinghouse

Index

Appendix

Allen-Bradley RSLinx

All online activities for Allen-Bradley products use a communications program called RSLinx. Before using the program to perform activities such as uploading, downloading or going online, RSLinx communications drivers need to be configured. To open RSLinx, you can either browse for it in the Programs list on your computer, or simply open the PLC program using RSLogix.

When a program is opened, RSLinx will often open automatically as a service. The icon farthest to the right in this picture is the RSLinx service. You can double click on the icon to open it, or browse for it in the Programs list.

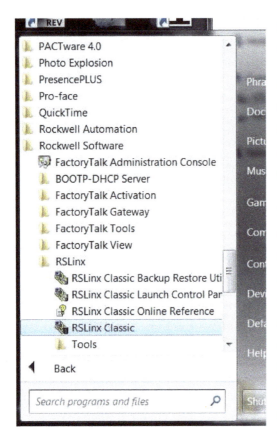

Drivers may already be configured in RSLinx. If they are not, the following procedures can be used.

Newer Allen-Bradley PLCs have several methods of communications, including USB, RS232 DF1 (Serial Communications) and Ethernet. USB communications is used with the ControlLogix L70 series of processors, the port is on the front of the processor.

Serial communications is available on all other processors. They either use a cable with a 9-pin plug on one end and a round plug on the other (Micrologix), or a 9-pin serial cable with a null modem adapter as shown in the RS-232 diagram in this document's Communications section. Ethernet uses a standard non-crossover Ethernet cable if using a switch (normal), or a crossover cable if attaching directly to the processor or Ethernet card. The communication ports are on the left side of the PLC for Micrologix, on the processor for the SLC 5/05 or CompactLogix, or on the bottom of the Ethernet card if using a ControlLogix system.

When a PLC is first commissioned, it has no Ethernet address assigned to it. The Ethernet address can either be assigned using the BootP Server utility from Allen-Bradley, or by downloading the program using the serial cable. If the Ethernet address has been set up in the PLC program, a serial download will also configure the Ethernet port.

To configure the serial driver:

1. Open RSLinx and select Communications > Configure Drivers.

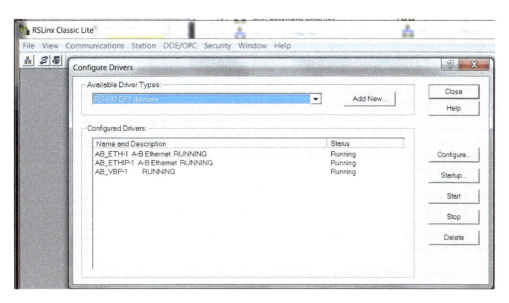

2. Select RS-232 DF1 devices in the pull-down menu under Available Driver Types. Press the "Add New" button. A dialog box will appear asking you to name the new driver. The default is AB_DF1-1, press the OK button to keep the default.

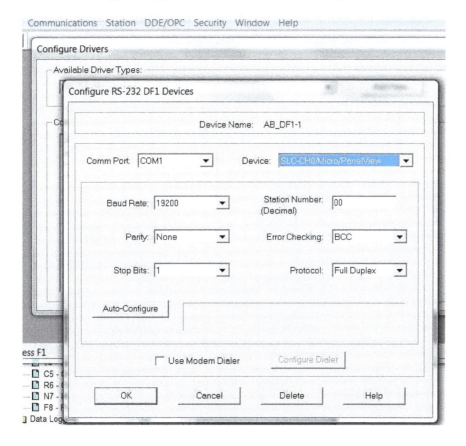

3. Under the Device pull-down, select SLC-CH0/Micro/Panelview (SLC or Micrologix) or Logix 5550/CompactLogix. Other settings are as shown in the above diagram except for Error Checking, which should be set to CRC.

If the cable is connected to the processor and it is powered up, you can press the Auto-Configure button. This should interrogate the PLC for its current settings and you will receive the message "Auto Configuration Successful!" This confirms that you have communications with the PLC and are ready for download.

Because most computers no longer have a serial port, it is often necessary to use a USB to Serial adapter. Some of these adapters will assign a serial port automatically; if so, you will need to use Device Manager on your computer to discover which port is used. If you see no assigned port for the adapter, try Port 8 or higher and press Auto Configure. Even if the driver displays "Port Conflict", it will usually show the "Auto Configuration Successful!" message anyway.

It is often necessary to use a "Null-Modem" adapter on the serial cable. This crosses pins 2 and 3 (TX and RX). This is true for ControlLogix and SLC500 processors.

To Configure the Ethernet Drivers:

There are two Ethernet drivers listed in the Available Driver Types list. The first is "Ethernet Devices" and the second is "Ethernet/IP Driver". **Ethernet Devices** allows the computer to locate processors and other devices by typing in the address; it works on all Allen-Bradley Ethernet items and also finds many compatible devices not made by Allen-Bradley.

Ethernet Devices:

1. Open RSLinx and select Communications > Configure Drivers.

2. Select Ethernet Devices in the pull-down menu under Available Driver Types. Press the "Add New" button. A dialog box will appear asking you to name the new driver. The default is AB_ETH-1, press the OK button to keep the default.

   If there is more than one group of Allen-Bradley processors and devices in your plant, it may be advisable to create more than one Ethernet Devices driver and give them different names, such as ETH_Line1, ETH_Line2 etc. This is to prevent the driver from attempting to find devices that are not present, which takes more time.

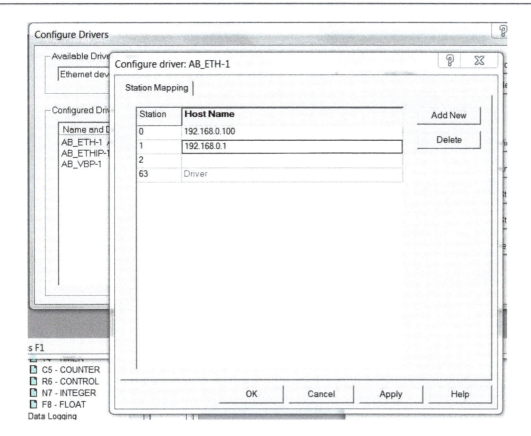

3. Type in the Ethernet addresses as shown. This assumes a computer with assigned address 192.168.0.100 and a PLC at 192.168.0.1. The first address (Station 0) is not necessary to communicate with the PLC, but it is there to show the current configuration of the computer itself. The second address (Station 1) is the address configured for the communications port in the PLC program. More PLCs and devices can be added to this list as required.

Ethernet/IP Driver:

This driver works for all Allen-Bradley CIP (Ethernet/IP) devices, which does not include the SLC 5/05 and Micrologix. It does not require that you know the addresses of the devices.

1. Open RSLinx and select Communications > Configure Drivers.

2. Select EtherNet/IP Driver in the pull-down menu under Available Driver Types. Press the "Add New" button. A dialog box will appear asking you to name the new driver. The default is AB_ETHIP-1, press the OK button to keep the default.

3. This driver only requires that you select your Ethernet card from the list of devices on your computer. This diagram show a wired port (192.168.4.204) and a wireless card (50.94.219.161)

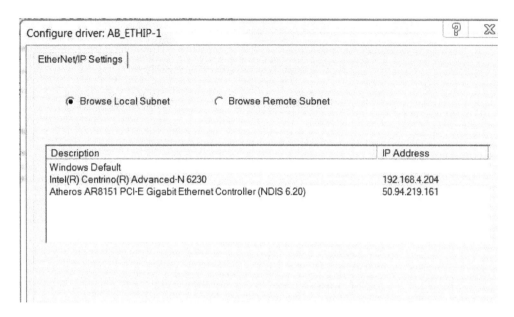

Since this driver does not maintain a list of devices, it is not necessary to create multiple drivers as with the Ethernet Devices driver. Previous connections may show up on the list when browsing in "RSWho", just right click on them and select "remove" to eliminate them.

Siemens Set PG-PC Interface

The setup for Siemens communications drivers can be found under Options in the Simatic Manager or in the block editor.

1.   Select "Set PG/PC Interface" to open the dialog.

2. There will be a number of drivers available for selection. For Ethernet, select the TCP/IP driver.

If Auto is selected, you will not be able to communicate through a bridge or router. If you are using a router, select the TCP/IP driver that ends with .1.

3. Clicking "Properties" allows modifications to the properties of the driver and to the network card (Network properties)

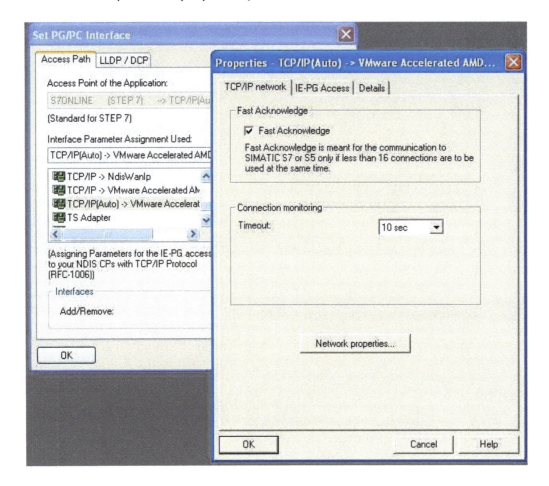

To see the processors available to connect to, select "Display Accessible Nodes" under the PLC tab. Siemens finds devices by MAC address rather than IP, so even devices outside of the Ethernet mask (domain) can be seen.

Other drivers that can be selected from the Set PG/PC Interface dialog include MPI (Serial RS232, 9 pin port) and Profibus (RS485). Both require a special adapter.

ASCII Tables

Table 1 - Basic ASCII

Dec	Hx	Oct	Char		Dec	Hx	Oct	Html	Chr	Dec	Hx	Oct	Html	Chr	Dec	Hx	Oct	Html	Chr
0	0	000	NUL	(null)	32	20	040	&#32;	Space	64	40	100	&#64;	@	96	60	140	&#96;	`
1	1	001	SOH	(start of heading)	33	21	041	&#33;	!	65	41	101	&#65;	A	97	61	141	&#97;	a
2	2	002	STX	(start of text)	34	22	042	"	"	66	42	102	&#66;	B	98	62	142	&#98;	b
3	3	003	ETX	(end of text)	35	23	043	&#35;	#	67	43	103	&#67;	C	99	63	143	&#99;	c
4	4	004	EOT	(end of transmission)	36	24	044	&#36;	$	68	44	104	&#68;	D	100	64	144	&#100;	d
5	5	005	ENQ	(enquiry)	37	25	045	&#37;	%	69	45	105	&#69;	E	101	65	145	&#101;	e
6	6	006	ACK	(acknowledge)	38	26	046	&	&	70	46	106	&#70;	F	102	66	146	&#102;	f
7	7	007	BEL	(bell)	39	27	047	'	'	71	47	107	&#71;	G	103	67	147	&#103;	g
8	8	010	BS	(backspace)	40	28	050	&#40;	(	72	48	110	&#72;	H	104	68	150	&#104;	h
9	9	011	TAB	(horizontal tab)	41	29	051	&#41;	)	73	49	111	&#73;	I	105	69	151	&#105;	i
10	A	012	LF	(NL line feed, new line)	42	2A	052	&#42;	*	74	4A	112	&#74;	J	106	6A	152	&#106;	j
11	B	013	VT	(vertical tab)	43	2B	053	&#43;	+	75	4B	113	&#75;	K	107	6B	153	&#107;	k
12	C	014	FF	(NP form feed, new page)	44	2C	054	&#44;	,	76	4C	114	&#76;	L	108	6C	154	&#108;	l
13	D	015	CR	(carriage return)	45	2D	055	&#45;	-	77	4D	115	&#77;	M	109	6D	155	&#109;	m
14	E	016	SO	(shift out)	46	2E	056	&#46;	.	78	4E	116	&#78;	N	110	6E	156	&#110;	n
15	F	017	SI	(shift in)	47	2F	057	&#47;	/	79	4F	117	&#79;	O	111	6F	157	&#111;	o
16	10	020	DLE	(data link escape)	48	30	060	&#48;	0	80	50	120	&#80;	P	112	70	160	&#112;	p
17	11	021	DC1	(device control 1)	49	31	061	&#49;	1	81	51	121	&#81;	Q	113	71	161	&#113;	q
18	12	022	DC2	(device control 2)	50	32	062	&#50;	2	82	52	122	&#82;	R	114	72	162	&#114;	r
19	13	023	DC3	(device control 3)	51	33	063	&#51;	3	83	53	123	&#83;	S	115	73	163	&#115;	s
20	14	024	DC4	(device control 4)	52	34	064	&#52;	4	84	54	124	&#84;	T	116	74	164	&#116;	t
21	15	025	NAK	(negative acknowledge)	53	35	065	&#53;	5	85	55	125	&#85;	U	117	75	165	&#117;	u
22	16	026	SYN	(synchronous idle)	54	36	066	&#54;	6	86	56	126	&#86;	V	118	76	166	&#118;	v
23	17	027	ETB	(end of trans. block)	55	37	067	&#55;	7	87	57	127	&#87;	W	119	77	167	&#119;	w
24	18	030	CAN	(cancel)	56	38	070	&#56;	8	88	58	130	&#88;	X	120	78	170	&#120;	x
25	19	031	EM	(end of medium)	57	39	071	&#57;	9	89	59	131	&#89;	Y	121	79	171	&#121;	y
26	1A	032	SUB	(substitute)	58	3A	072	&#58;	:	90	5A	132	&#90;	Z	122	7A	172	&#122;	z
27	1B	033	ESC	(escape)	59	3B	073	&#59;	;	91	5B	133	&#91;	[	123	7B	173	&#123;	{
28	1C	034	FS	(file separator)	60	3C	074	&#60;	<	92	5C	134	&#92;	\	124	7C	174	&#124;		
29	1D	035	GS	(group separator)	61	3D	075	&#61;	=	93	5D	135	&#93;	]	125	7D	175	&#125;	}
30	1E	036	RS	(record separator)	62	3E	076	&#62;	>	94	5E	136	&#94;	^	126	7E	176	&#126;	~
31	1F	037	US	(unit separator)	63	3F	077	&#63;	?	95	5F	137	&#95;	_	127	7F	177	&#127;	DEL

Source: www.LookupTables.com

Table 2 - Extended ASCII

128	Ç	144	É	160	á	176	▒	192	└	208	╨	224	α	240	≡
129	ü	145	æ	161	í	177	▓	193	┴	209	╤	225	ß	241	±
130	é	146	Æ	162	ó	178	█	194	┬	210	╥	226	Γ	242	≥
131	â	147	ô	163	ú	179	│	195	├	211	╙	227	π	243	≤
132	ä	148	ö	164	ñ	180	┤	196	─	212	╘	228	Σ	244	⌠
133	à	149	ò	165	Ñ	181	╡	197	┼	213	╒	229	σ	245	⌡
134	å	150	û	166	ª	182	╢	198	╞	214	╓	230	µ	246	÷
135	ç	151	ù	167	º	183	╖	199	╟	215	╫	231	τ	247	≈
136	ê	152	ÿ	168	¿	184	╕	200	╚	216	╪	232	Φ	248	°
137	ë	153	Ö	169	⌐	185	╣	201	╔	217	┘	233	Θ	249	·
138	è	154	Ü	170	¬	186	║	202	╩	218	┌	234	Ω	250	·
139	ï	155	¢	171	½	187	╗	203	╦	219	█	235	δ	251	√
140	î	156	£	172	¼	188	╝	204	╠	220	▄	236	∞	252	ⁿ
141	ì	157	¥	173	¡	189	╜	205	═	221	▌	237	φ	253	²
142	Ä	158	₧	174	«	190	╛	206	╬	222	▐	238	ε	254	■
143	Å	159	ƒ	175	»	191	┐	207	╧	223	▀	239	∩	255	

Source: www.LookupTables.com

Exercise Solutions

Exercise 1 p.38

Q5 –   The Control level.

Exercise 2 p. 47

Q1 –   Decimal: 27,831   Hexadecimal: 6CB7   Octal: 66,267
This number cannot be converted to BCD, two of the binary groups are higher than 1001.

Q2 –   Signed Integer: -27,255

Q3 –   BCD: 0100_0001_0111 = 417   Binary: 1 1010 0001   Hexadecimal: 1A1

Q4 –   Binary: 10 1010 1001 1110   Decimal: 10,910

Q5 –   There are 4 Bytes in a Double Integer

Q6 –   Some brands of PLC call a Tag a Symbol, but symbols usually are a shortcut to a numerical data address register.

Exercise 3 p. 54

Q1 –   Decimal, Octal and Hexadecimal.

Q2 –   The Main Routine

Q3 –   The hardware

Q4 –   Modules, communication ports, I/O networks, addresses

Exercise 4 p. 64

Q1 –   Instruction List (IL), Ladder (LAD), Function Block Diagram (FBD), Structured Text (ST), Sequential Function Charts (SFC)

Q2 –   1. Read physical inputs to Input Image Table. 2. Solve Logic 3. Write Output Image Table to physical outputs. 4. "Housekeeping" tasks.

Q3 –   Yes, online editing.

Exercise 5 p. 69

Q1 -

Q2 -

Exercise 6 p. 74

Q1 -

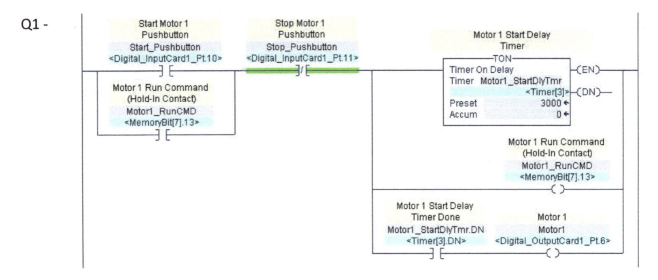

Q2 –

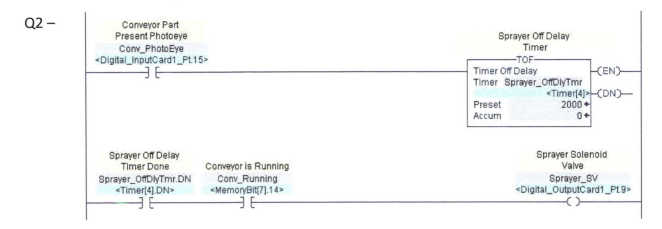

Exercise 6 p. 74 continued

Q3 –

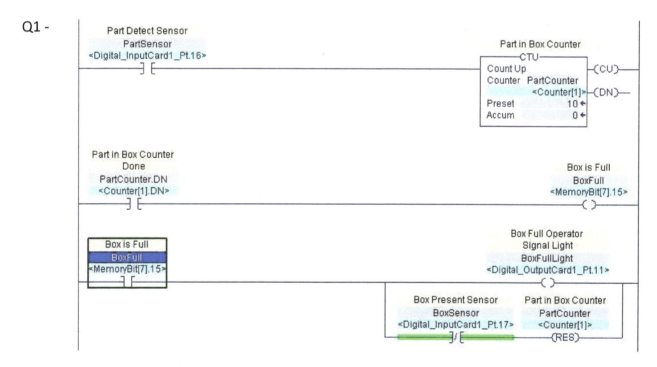

Exercise 7 p. 78

Q1 -

Exercise 7 p. 78 continued

Q2 –

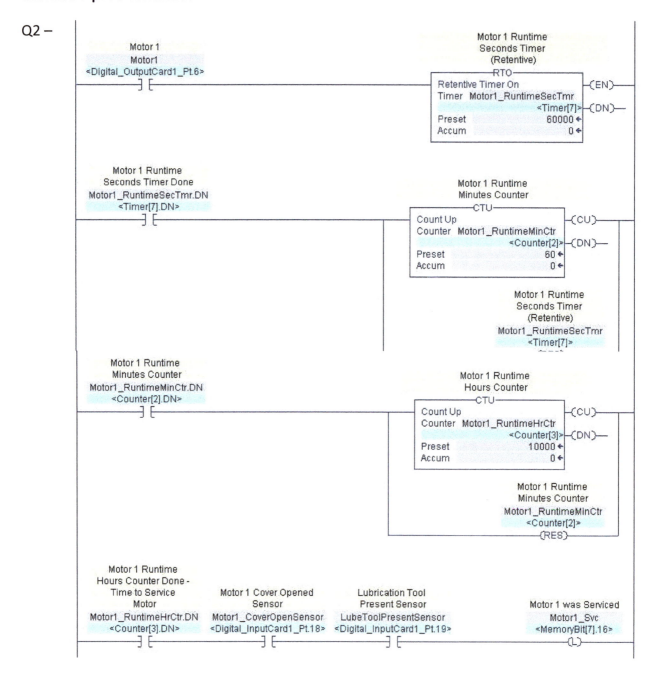

Exercise 7 p. 78 (Q2) continued

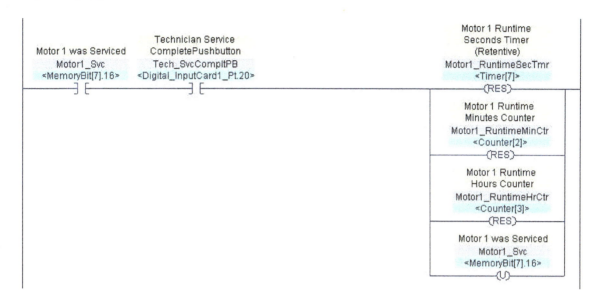

Exercise 8 p. 84

Q1 -

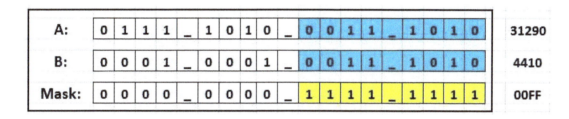

A:	0	1	1	1	_	1	0	1	0	_	0	0	1	1	_	1 0 1 0	31290
B:	0	0	0	1	_	0	0	0	1	_	0	0	1	1	_	1 0 1 0	4410
Mask:	0	0	0	0	_	0	0	0	0	_	1	1	1	1	_	1 1 1 1	00FF

Yes, they are equal through the mask.

Exercise 8 p. 84 Continued

Q2 –

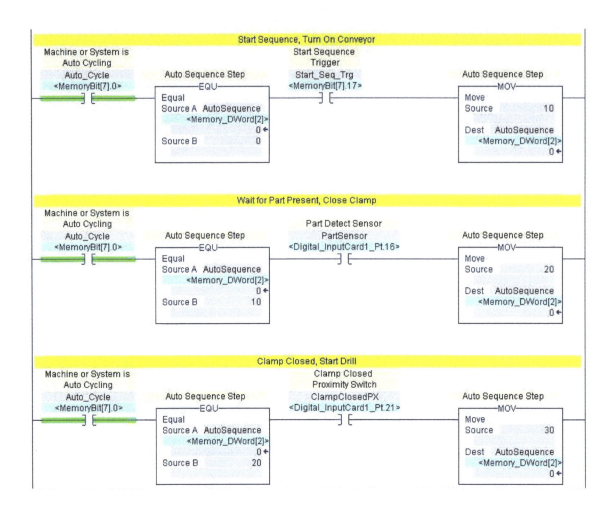

Auto sequences increment by 10 so that an extra step (i.e. Step 15) can be inserted if necessary.

Exercise 9 p. 90

Q1 -   1750/31,760 = 0.0551 (Scaling Factor)

Test: 31,760 * 0.0551 = 1749.976. Close enough! So, <Sensor Value> * 0.0551 = RPM.

Percent = RPM/Max RPM * 100%

(Continued next page)

Exercise 9 p. 90 Q1 Continued

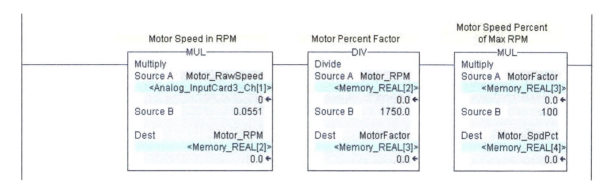

Q2 –

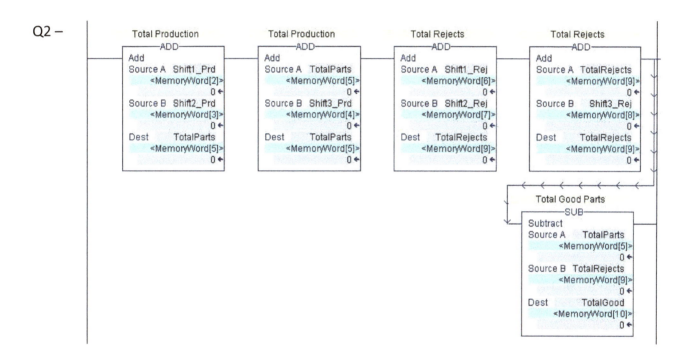

Exercise 10 p. 93

Q1 –   M = (Y2-Y1)/(X2-X1) = (6000-0)/(24780-96) = 0.24307

B = Y1 – (M * X1) = 0 – (0.24307 * 96) = -23.33472

Liters = Gallons * 3.78541

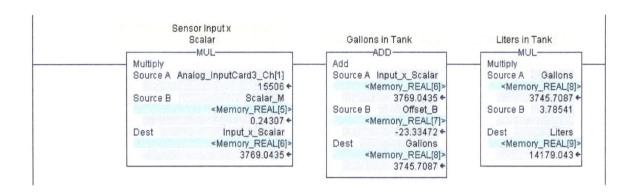

Exercise 11 p. 101

Q1 –   Trigonometric functions are usually used in motion control and positioning applications, they calculate geometrical coordinates.

Q2 -   47 _G_ 6F _o_ 6F _o_ 64 _d_ 20 ___ 4A _J_ 6F _o_ 62 _b_ 21 _!_
        Good Job!

Q3 -   Yes, Jump instructions can jump backward. If so, a method to exit the loop is necessary, such as incrementing a counter.

Q4 -   FIFO: First In First Out   LIFO: Last In First Out

Q5 -   A Sequencer monitors and controls repeatable operations.

Exercise 12 p. 107

Q1 -   False

Q2 -   True

Q3 -   Q98.5 "PP04" will be off.

Q4 -   Q98.5 will be on.

Q5 -   Look for Coils and operations affecting digital and analog outputs.

Q6 -   Physical inputs, and contacts with no coil.

Photo Credits

Judicious use of Wikipedia, PLCS.net, MrPLC.com, PLCDEV.com and various other websites have also been employed in this research.

p.6: "Koyo DL405" by Automation Direct. Digital image. www.automationdirect.com. Retrieved Oct. 1, 2016.

p.6: "Modicon Premium" by Schneider Electric. Digital image. www.schneider-electric.com. Retrieved Oct. 1, 2016.

p.6: "Omron CJ1" by Omron. Digital image. www.ia.omron.com. Retrieved Oct. 1, 2016.

p.9: "Harvard Mark I constant switches detail" by Arnold Reinhold (Own work) [CC BY-SA 3.0 (http://creativecommons.org/licenses/by-sa/3.0) or GFDL (http://www.gnu.org/copyleft/fdl.html)], via Wikimedia Commons

p.9: "Harvard Mark I program tape" by Arnold Reinhold (Own work) [CC BY-SA 3.0 (http://creativecommons.org/licenses/by-sa/3.0) or GFDL (http://www.gnu.org/copyleft/fdl.html)], via Wikimedia Commons

p.10: "ENIAC_Penn2" by TexasDex at English Wikipedia [GFDL (http://www.gnu.org/copyleft/fdl.html) or CC-BY-SA-3.0 (http://creativecommons.org/licenses/by-sa/3.0/)], via Wikimedia Commons

p.11: "KL Core Memory" by Konstantin Lanzet (received per Email Camera: Canon EOS 400D) [GFDL (http://www.gnu.org/copyleft/fdl.html) or CC-BY-SA-3.0 (http://creativecommons.org/licenses/by-sa/3.0/)], via Wikimedia Commons

p.15: "Development team" by Segal0577 (Own work) [CC BY-SA 4.0 (http://creativecommons.org/licenses/by-sa/4.0)], via Wikimedia Commons

p.16: "Mitsubishi Handheld Programmer F1-20P" by Industry Marketplace. Digital image. www.industry-marketplace.com. Retrieved Nov. 15, 2016

p.16: "Allen Bradley Bulletin 1774 PLC" by Concepts Industrial Asset Disposal, Inc. Retrieved Oct. 1, 2016

p.23: "Pushbutton" by Automation Direct. Digital image. www.automationdirect.com. Retrieved Oct. 1, 2016.

p.23: "Photoeye" by Automation Direct. Digital image. www.automationdirect.com. Retrieved Oct. 1, 2016.

p.23: "Proximity Switch" by Automation Direct. Digital image. www.automationdirect.com. Retrieved Oct. 1, 2016.

p.23: "Pilot Light" by Automation Direct. Digital image. www.automationdirect.com. Retrieved Oct. 1, 2016.

p.23: "Motor Starter" by Automation Direct. Digital image. www.automationdirect.com. Retrieved Oct. 1, 2016.

p.23: "Solenoid Valve" by Automation Direct. Digital image. www.automationdirect.com. Retrieved Oct. 1, 2016.

p.24: "Electromechanical Relay" by Automation Direct. Digital image. www.automationdirect.com. Retrieved Oct. 1, 2016.

p.25: "Potentiometer" by Iainf - Self-photographed, CC BY 2.5, https://commons.wikimedia.org/w/index.php?curid=1407612

p.25: "Pressure Transducer" by Automation Direct. Digital image. www.automationdirect.com. Retrieved Oct. 1, 2016.

p.25: "Platinum RTD" by Automation Direct. Digital image. www.automationdirect.com. Retrieved Oct. 1, 2016.

p.25: "Linear Actuator" by Automation Direct. Digital image. www.automationdirect.com. Retrieved Oct. 1, 2016.

p.25: "VFD Speed" by Automation Direct. Digital image. www.automationdirect.com. Retrieved Oct. 1, 2016.

p.34: "Twisted Pair" by Spinningspark at en.wikipedia [CC BY-SA 3.0 (http://creativecommons.org/licenses/by-sa/3.0)], via Wikimedia Commons

p.36: "Traditional serial RS485 network for Accuenergy Multifunction Power and Energy Meters" by Accuenergy (Canada) Inc. Digital Image. www.accuenergy.com.

p.42: "Thumbwheel Switch". Digital image. www.ia.omron.com. Omron. Retrieved Oct. 1, 2016.

NTH University is an engineering training program for interns, apprentices, and employees of Automation NTH (www.automationnth.com). Automation NTH is a control systems integrator located near Nashville, Tennessee. They also provide turnkey manufacturing and machinery solutions through their Certified Automation Partner (CAP) program.

About the Author

Frank Lamb is an industrial automation consultant and advanced PLC programming trainer with more than 30 years of experience in controls and machine automation.

From 1996 to 2006, Frank owned and operated Automation Consulting Services, Inc. (ACS), a panel building and machine integration company in Knoxville, TN. From 2006 to 2011, he worked as a senior-level project engineer for Wright Industries in Nashville, TN, where he led the design and implementation of large, complex systems and custom machines for multinational corporations and government agencies. In December 2011, Frank re-established Automation Consulting, LLC with a new vision: to use his experience in the field of industrial automation – from electrical, mechanical and controls engineering to project management, training, and machine documentation – to provide expert consulting and training services to manufacturers.

Frank is the president and owner of Automation Consulting, LLC in Nashville and works as Lead Trainer for Automation NTH in LaVergne, Tennessee. He is the author of several books including *Industrial Automation: Hands On*, published by McGraw-Hill Professional in 2013, *Advanced PLC Hardware and Programming*, published by Automation Consulting, LLC in 2019, and *Maintenance and Troubleshooting in Industrial Automation*, published by Automation Consulting, LLC in 2022. He is a United States Air Force veteran, received his BSEE in Electrical and Computer Engineering from the University of Tennessee, and has a Green Belt in Lean Manufacturing/Six Sigma from Purdue University.